4G Roadmap and Emerging Communication Technologies

For a complete listing of the *Artech House Universal Personal Communications series*, turn to the back of this book.

4G Roadmap and Emerging Communication Technologies

Young Kyun Kim
Ramjee Prasad

ARTECH
HOUSE

BOSTON | LONDON
artechhouse.com

Library of Congress Cataloging-in-Publication Data
Kim, Young Kyun.
    4G wireless roadmap and emerging technologies/ Young Kyun Kim, Ramjee Prasad.
    p. cm.
    Includes bibliographical references and index.
    ISBN 1-58053-931-9 (alk. paper)
    1. Wireless communication systems—Technological innovations.   2. Cellular telephone
systems—Technological innovations.   3. Wireless LANs—Technological innovations.   I.
Prasad, Ramjee.   II. Title
    TK5103.2.K56   2005
    621.384—dc21   2005055857

British Library Cataloguing in Publication Data
Kim, Young Kyun
    4G wireless roadmap and emerging technologies. — (Artech House universal personal com-
munications series)
    1. Wireless communication systems   I. Title   II. Prasad, Ramjee   III. Four G
    621.3'821

    ISBN–10: 1-58053-931-9

Cover design by Leslie Genser

International Standard Book Number: 1-58053-931-9
Library of Congress Catalog Card Number: 2005055857

10 9 8 7 6 5 4 3 2 1

To my wife Myung Hee, to our son and daughter-in-law Jee Whan and Jane, and to our daughter Soyoung
—Young Kim

To my wife Jyoti, to our daughter Neeli, to our sons Anand and Rajeev, and to our grandchildren Sneha, Ruchika, and Akash
—Ramjee Prasad

Open mind and cooperative spirit are the starting point of globalization
Kun-Hee Lee
Chairman of Samsung Group

You don't predict the future and then wait. You create the future.
Jong-Yong Yun
CEO of Samsung Electronics

Contents

Preface

गतसंगस्य मुक्तस्य ज्ञानावस्थितचेतसः ।
यज्ञायाचरतः कर्म समग्रं प्रविलीयते ॥२३॥

gata-saṅgasya muktasya
jñānāvasthita-cetasaḥ
yajñāyācarataḥ karma
samagraṁ pravilīyate

The work of a man who is unattached to the modes of material nature and who
is fully situated in transcendental knowledge merges entirely into transcendence.
—Bhagavad Gita (4.23)

During the Samsung 4G Forum and advisory meetings at Samsung Electronics (Korea) and CTIF (Denmark) in 2003 and 2004, we discussed about the evident shortage of information on 4G for both academics and industry. This discussion motivated us to write a book on 4G that will help the researchers and engineers to understand the fundamental nature of future 4G wireless communication systems and will answer the questions: What is 4G and where it will lead us? As 4G is rapidly becoming the center of attention of industry and academia, we authors, representing both groups, approached this vast subject from a wide perspective, aiming to provide the reader with a complete and timely vision of future communication systems. We live in an exciting time, the foundations of 4G are being discussed and developed, and huge international aim to find a global consensus on 4G. Attempts to define 4G abound, but paradoxically, it is hard to find a widely accepted universal definition of 4G. Along this book we will approach 4G from several directions, striving for clarity and openness in the discussions.

This book addresses in a comprehensive and balanced way the major driving forces behind 4G systems, including scenarios, enabling technologies, and applications. Roadmaps, expectations, and visions for 4G systems are also discussed. Background information and the evolutionary development of wireless technologies for future 4G networks are considered in detail. The book also goes deeper into the technical difficulties and challenges of 4G implementation. In addition to dealing with the fundamental aspects of the air interfaces and networking, the reader gets a complete panorama of other 4G related issues, including key technologies, spectrum, standardization, ongoing worldwide research cooperation projects, and the impact of 4G on future communications equipment, in particular 4G terminals. It is the first book to discuss many of the challenges in the advancement of 4G, including the technical and practical realities of the progression to 4G. We wrote this book for the following types of readers: technologists, engineers, scientists, lecturers and professors, and Ph.D. candidates; and with the objective to have a single source reference that offers detailed information about 4G on the topics as shown in Figure P.1.

Chapter 1 is an introduction to 4G, motivating its development and sketching some initial visions and definitions. Chapter 2 provides a deeper

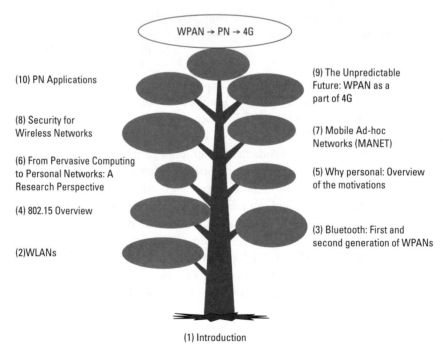

Figure P.1 Coverage of the book.

insight into future 4G systems, starting with a user-centric approach aiming to define 4G scenarios and services and ending with a series of realistic 4G visions, including related 4G spectrum issues as well. Chapter 3 discusses multicarrier-based multiple access techniques, describing the most promising concepts likely to be used in 4G. The most important enabling technologies for 4G are presented and discussed in Chapter 4. These include multiple antenna techniques, radio resource management, software radio technology, mobile IP and relaying techniques. Chapter 5 gives a broad overview of the major research and development activities concentrated on 4G. Chapter 6 discusses 4G terminals from technology and applications standpoints. Finally, Chapter 7 wraps up the ideas and concepts discussed through the book, presenting appealing 4G visions of the near and distant future.

We have tried our best to make each chapter complete in itself. This book will give successful direction to researchers and engineers working in universities, research laboratories, and industry for making 4G happen. We cannot claim that the book is errorless. Any remarks to improve the text and correct any errors would be greatly appreciated.

Young Kyun Kim, Suwon, Korea
Ramjee Prasad, Aalborg, Denmark
October 2005

Acknowledgments

We would like to express our hearty appreciation to Dr. Marcos Katz of Samsung Electronics, Korea, for his invaluable contribution to this book. The contribution he gave had confirmed his deep commitment to scientific and technical matters, his professional capability, and, most importantly, his enthusiastic approach to solve complex problems. We would not have completed this book without his devoted support. In thanking him we would like to take this opportunity to wish him all the success in his future career that he fully deserves. We would also like to thank the efforts of our colleagues from Samsung Electronics, Korea, in particular the contributions of Junhwan Kim, Euntaek Lim and Juyeon Song in Chapter 2, Chanbyoung Chae and Seokhyun Yoon in Chapter 4 and Bosun Jung, Wuk Kim, Sangkyung Sung, Jaekwon Oh, Kyung-Tak Lee, and Byung-Rae Lee in Chapter 6. In addition, we thank our colleagues from Samsung Electronics Research Institute (UK) Terence Dodgson and Byron Bakaimis for their contribution in Chapter 4. We would specially like to thank Mr. Jong-Yong Yun, CEO of Samsung Electronics and Mr. Ki-Tae Lee, President of Samsung Electronics Telecom Business for their 4G leadership and consistent support in global 4G research activities.

We wish to thank many other colleagues from Samsung Electronics in Korea as well as Center for TeleInFrastruktur (CTIF) in Denmark who supported us in finishing this book. Finally we appreciate the support of Mrs. Junko Prasad in completing this book.

1

Introduction

4G is characterized by three words: ubiquitous, mobile, and broadband.
— 3rd Samsung 4G Forum,
Jeju Island, Korea, August 2005

4G shall explore more on application level using available networks and should be a Mediator of all the existing and emerging networks that should be easily accessible by everybody wherever, whenever, whatever.
—First 4G International Workshop,
New Delhi, India, November 2004

Since the inception of mobile communications in the early 1980s, we have witnessed an ever-growing increase in the development of mobile communication technology. Analog wireless communication systems were replaced by digital ones, voice services are being complemented with data services, supported data transfer speeds have increased by more than a thousand-fold, network coverage has been stretched to cover virtually entire countries and continents, and many other remarkable achievements have taken place in a relatively short period. Many visions, even the most optimistic ones, failed to foresee that in just a quarter century the number of mobile subscribers will approach one-third of the world's population and surpass wireline subscribers. Unveiling the future of wireless systems was then perhaps a more difficult task than it would be today, since now we can expect that part of the future wireless networks will be the result of an evolution of the current networks. Establishing a sound vision of future mobile communications systems is one of the main purposes of this book. Large portions of this book will be devoted to the present and discuss future 4G networks from different perspectives [1-110].

1

Presently, as third generation (3G) International Mobile Telecommunications 2000 (IMT-2000) systems are being deployed, further developments aiming at their enhancement are being conducted on a worldwide scale. These research and development activities surrounding Beyond 3G (B3G) systems target not only the task of creating a universal vision of future mobile communications systems but also the identification and further evaluation of technical solutions for realizing such systems. All these initiatives are mainly driven by the ever-increasing growth in the number of subscribers as well as by users' demands of high-speed connectivity on the move. The demand of ubiquitous multimedia services in a personalized fashion is the most salient requirement of future users. When this is seen from the perspective of millions of users, one can easily see that such broadband services will require the integration of several wireless networks of different types, including networks supporting much higher data rates than what we are accustomed to experience today. In the broadest sense, the fourth generation (4G) networks will integrate technologies from broadcasting networks to wide area and metropolitan networks down to smaller networks like wireless local and personal area networks, all under the umbrella of a single, monolithic network: the 4G network. Data rates are really what broadband is about. It is expected that broadband wireless communications will support applications up to 1 Gbps. We can say that two major paths will conduct us towards 4G, namely, through *evolution* of current systems, and through *revolution* by creating new systems able to provide very high data throughputs. These two approaches will coexist and be complementary. In this visionary perspective of the road ahead, in order to keep pace with society's communication needs in the years to come, capacity will be one of the major issues to be developed due to the foreseen increase in demand for new services (especially those based on multimedia). Together with this, personal mobility will impose new challenges to the development of new personal and mobile communications systems.

A conclusion can be drawn from this: Even if at a certain point it may appear academic to develop a system for a capacity much higher than what seems reasonable (in the sense that it is not straightforward to identify applications requiring such high capacity), it is still worthwhile since almost certainly future applications will come out that need a capacity of 1 Gbps or even more. The story of fiber optics plays this out. Moreover, high capacity networks will reduce service costs dramatically; helping 4G to get a solid grip after it is launched. Higher data rates can be also justified from the user terminal standpoint. Indeed, as advanced terminals with audio and imaging capabilities are becoming more and more common, need for much higher network data throughput capabilities will arise due to the expected enormous increase of traffic in both uplink and downlink. Rapid development will shrink the world into a global information multimedia communication village (GIMCV) by 2020. Figure 1.1 illustrates the basic concept of a GIMCV, which consists of version

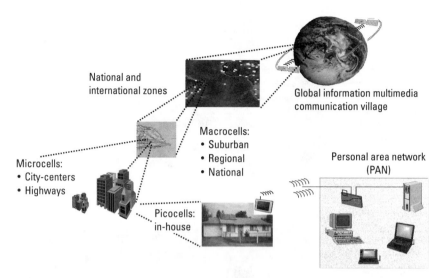

Figure 1.1 Global information multimedia communication village.

components of different scales ranging from global to pico-cellular in size. Figure 1.2 shows a family tree of the GIMCV system [1].

The passage from generation to generation is not only characterized by an increase in the data rate, but also by the transition from pure circuit switched (CS) systems to CS-voice/packed data and IP-core-based systems, as is highlighted in Figure 1.3 [1-94].

1.1 Visions

Economic and technical trends together with applications requirements will drive the future of mobile communications. As will be discussed in the next chapter, the increase of the mobile subscriber base will be accompanied by a sharp increase of data usage. Ultimately, it is expected that the volume of data traffic will substantially exceed that of voice traffic. Mobile Internet, one of the main drivers for multimedia applications, will speed up this process.

The number of mobile subscribers has increased much faster than expected and it will continue to grow through the 2000s. The universal mobile telecommunications system (UMTS) Forum expects that in Europe more than 90 million mobile subscribers will use mobile multimedia services in 2010, with a data traffic share of about 60% of the total traffic. Worldwide mobile Internet subscribers are expected to top half a billion in the same year, according to the same source. Even higher figures are expected in Asia. Additional frequency assignment will be necessary for 3G to accommodate the growing demand. The bandwidth to be added is assumed to be of several hundred megaherts in 2010.

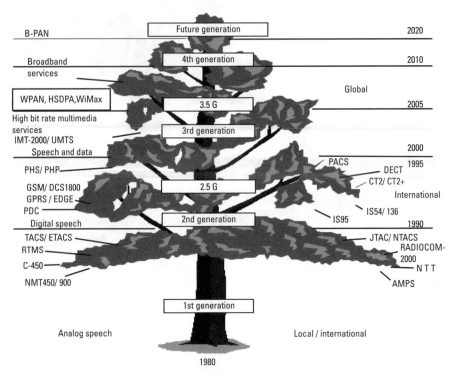

Figure 1.2 Family tree of the GIMCV. Branches and leaves of the GIMCV family tree are not shown in chronological order.

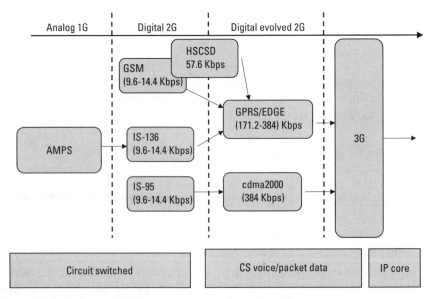

Figure 1.3 Evolution of cellular communications from 2G to 3G.

However, the added bandwidth greatly depends on the growth ratio of traffic per subscribers. Therefore, study of high capacity cellular systems with improved spectrum efficiency and new bands is necessary to accommodate growing traffic in 2010 and beyond. Higher data rates and wireless Internet access are key components of the future mobile communications systems. They are also key concepts in 3G systems as well. Future mobile communications systems should bring something more than just faster data or wireless Internet access [28, 57, 58]. Something that we are missing today, even in 3G, is the flexible and seamless interoperability of various existing networks like cellular, cordless, wireless local area network (WLAN) type systems, systems for short connectivity, and wired systems. One of the greatest challenges for 4G is to integrate the whole worldwide communication infrastructure to form one transparent network allowing various and different access systems to connect into it depending on the user's needs, location, and network availability. The heterogeneity of various accesses is one of the principal distinctive aspects of future networks. The different access systems are organized in a layered structure according to the application areas, cell ranges, and radio environments. This allows a flexible and scalable environment for system deployment [110].

The creation of this new network requires a completely new design approach. So far, most of the existing systems have been designed in isolation without taking into account a possible interworking with other access technologies. Their system design is mainly based on the traditional vertical approach to support a certain set of services with a particular technology. The universal terrestrial radio access (UTRA) concept has already combined the frequency division multiplexing (FDD) and time division multiplexing (TDD) components to support the different symmetrical and asymmetrical service needs in a spectrum-efficient way. This is the first step to a more horizontal approach [59, 39] where different access technologies will be combined into a common platform to complement each other in an optimum way for different service requirements and radio environments. Because of the dominant role of Internet Protocol (IP)-based data traffic, these access systems will be connected to a common, flexible and seamless IP-based core network. This will result in a lower infrastructure cost, faster provisioning of new features, and easy integration of new network elements.

This network could be supported by technologies like JAVA Virtual Machines and CORBA. The vision of the seamless future network is shown in Figure 1.4.

The mobility management will be part of a new Media Access System, which serves as an interface between the core network and the particular access technology to connect a user via a single number for different access systems to the network. Global roaming for all access technologies is required. The internetworking between these different access systems in terms of horizontal

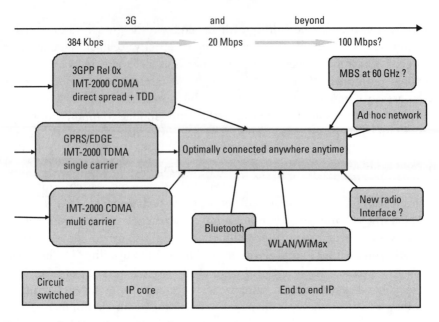

Figure 1.4 Seamless future networks.

and vertical handover and seamless services with service negotiation with respect to mobility, security and quality of service (QoS), will be a key requirement. The latter will be handled in the common Media Access System and the core network. Multimode terminals and new appliances are also key components to support these different access technologies of the common platform seamlessly from the user perspective. These terminals may be adaptive, based on high signal processing power. Therefore, the concept of software defined radio, in which different access technologies can be supported by automatic adaptation to any available network technologies at any particular instance could be a key technology in the future perspective.

In order to realize the vision of systems beyond third generation, many technical challenges have to be solved. In spite of the 2-Mbps data rates achievable by 3G systems, the end user throughput of these systems could still be only a small fraction of the actual need of truly broadband systems, taking into consideration economic and practical efficiency. Scarcity of available radio spectrum still remains a bottleneck, and researchers worldwide are working on different spectrum-efficient techniques that can adapt to varying environmental conditions. The common aim is to exploit effectively the limited radio resources [23, 24], taking into account the interaction among different layers of the protocol stack. The best efficiency is achieved if the system is able to adapt to environmental conditions that change in time, location, and even available service mixture. This leads to the investigation of adaptive radio interface and network systems where diversity means [60-69], coding method as well as type and gain [70-72], modulation method or order (M-QAM, M-PSK, multicode/single code, multicarrier/single carrier) [73-75], data rate, data packet size, channel allocation, and service selection can be adapted effectively. New radio access concepts [76], advanced reception techniques [77-82], smart antenna technology [63-72], and resource management issues [83-90] are the main research areas that could realize the major goals of future generation wireless systems [91].

1.2 Next Generation

Next generation, commonly known as the fourth generation, will mainly focus on wireless IP with self-provisioning of different multimedia services such as audio, video, and games. Note that all the considered access technologies overlap each other to some extent in some areas of services, as illustrated in Figure 1.5 Thus, a move towards integration of technology is a logical next step to provide service continuity and higher user experience (quality of experience), as suggested by Figure 1.6.

The ITU-R vision of future wireless communication systems also calls for integration of technologies, which are commonly known as heterogeneous systems, B3G systems, or beyond IMT-2000 (as described by the ITU) systems (to some people B3G could mean any standard or technology developed after 3G). Technology integration will provide adequate services to a user depending on mobility and availability. Of course, this involves several new challenges; for example, handover/handoff or mobility, security, QoS, and cost efficiency. These issues should be resolved without changing the existing standards.

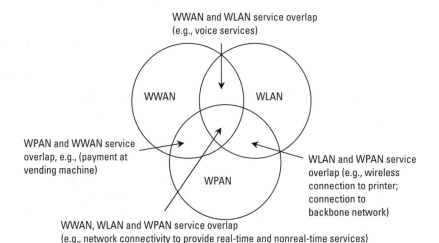

Figure 1.5 WWAN, WPAN, and WLAN overlap.

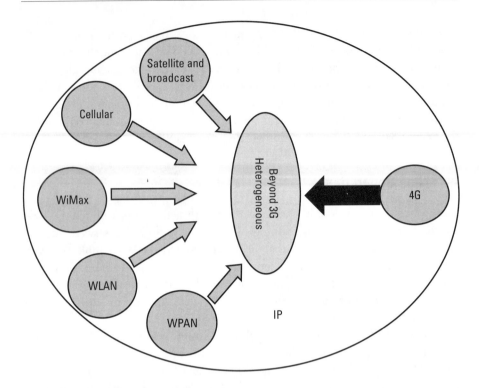

Figure 1.6 Future of telecommunications.

Seamless handover should be provided while a user moves from the network of one access technology to the other and from the domain of one stakeholder to the other. Seamless handover means the provision of seamless service while the user is mobile (i.e., the user does not perceive any disruption in service or quality even during handover).

The ITU-R vision also refers to a new air interface, also known as 4G. Since any new system takes about 10 years to develop and deploy (see Figure 1.7), work on B3G and 4G has already started; a possible solution is given in [91]. The current market shows that 3G is being deployed and adopted slower than expected, and hopefully some lessons will be learned from it while developing 4G [93].

As already mentioned, we expect that future 4G systems will develop in two ways, namely evolution and revolution, and once deployed, legacy networks, their evolution, and newly developed networks will coexist under the umbrella of 4G [110]. From this standpoint, 4G has a integrative role, and a 4G network can be seen as a convergence platform encompassing highly heterogeneous networks [111].

A possible future scenario is given in Figure 1.8, which shows all technologies working together while providing all the services to the users anywhere and

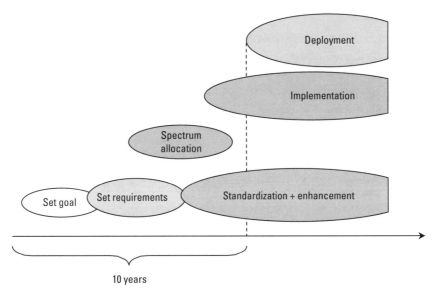

Figure 1.7 Time required for new technology development and deployment.

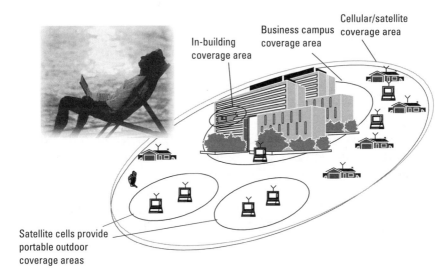

Figure 1.8 Future of wireless.

anytime. Table 1.1 shows the envisaged development of stakeholders of various networks and technological development for short-, mid- and long-term future. The table also points out several technological issues that should be worked on. Arrows between two cells of the table show the possibility of handover between

Table 1.1
Envisaged Technology Development in Short, Mid, and Long Term

Stakeholder (for handover; of one or more access networks)	IP	Broadcast (DVB-T, DAB, etc.)	WWAN (3G, 2.5G, etc.)	WLAN (IEEE 802.11)	WiMax	WPAN and Ad Hoc (IEEE 802.15, etc.)
Short term (2 to 3 years)	Same (not broadcast)	v4	Similar to TV and radio	3G and 2.5G, handover: maybe	b, g, a, n, s, MAC enh.	WiMax	
Mid-term (3 to 5 years)	Same (maybe few different, surely not broadcast)	v4 and 6	As above	3G 3.5G and 2.5G, handover: possible	g, a, NG QoS etc.	WiMax, MBWA	
Long term (5 to 10 years)	Same and different	v4 and 6	SDR	2.5G, 3G, 3.5G, and 4G, handover: must	g, a, n, s, NG QoS etc.	WiMax, MBWA, NG	UbiCom

NG (+) Next Generation and beyond
The arrows indicate handover operation. In the table, the darker the arrow, the more common will be the handovers between the concerned technologies.

the two technologies, and the shade of the arrow (grayscale) shows the expected extent of the handover. Research work should be done on seamless handover, which involves the study of several issues like security and QoS, which should be done at each protocol layer and network element. This topic itself will require further study on development methods and technologies including hardware, software, and firmware, and technologies like application specific integrated circuits (ASICs). Another important research topic is software defined radio (SDR), which includes reconfigurability at every protocol layer.

WLANs provide roaming within LANs, and work is happening towards further enhancement in this field. While wireless wide area networks (WWANs) provide roaming too, the challenge now is to provide seamless roaming from one system to another, from one location to another, and from one network provider to another. In terms of security, both WLANs and WWANs have their own approach. The challenge is to provide the level of security required by the user while roaming from one system to another. The user must get end-to-end security independent of any system, service provider, or location. Security also incorporates user authentication, which can be related to another important issue: billing. Both security and roaming must be based on the kind of service a user is accessing. The required QoS must be maintained when a user roams from one system to other. Besides maintaining the QoS it should be possible to know the kind of service that can be provided by a particular system, service provider, and location. Work on integration of the WLANs and the wireless personal area network (WPANs) must also be done. The biggest technical challenge here will be the coexistence of the two devices, as both of them work in the same frequency band.

Another area of research for the next generation communications will be in the field of personal networks (PNs) [94]. PN provides a virtual space for users that spans a variety of infrastructure technologies and ad hoc networks. In other words, PNs provide a personal distributed environment where people interact with various companions, embedded or invisible computers, not only in their vicinity but potentially anywhere. Figure 1.9 portrays the concept of PNs. Several technical challenges arise with PNs-besides interworking between different technologies-some of which are security, self-organization, service discovery, and resource discovery [94].

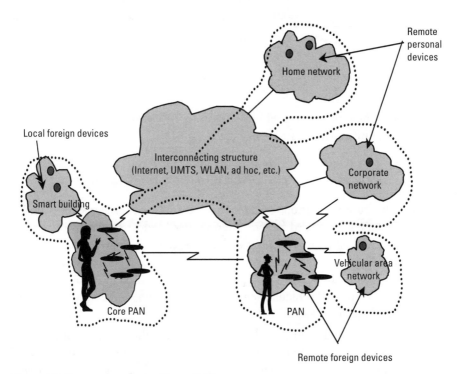

Figure 1.9 Personal networks. (From: [94].)

1.3   What Is 4G?

Answering this question is not by any means a simple task. It depends on who (manufacturers, operators, academia, regulatory bodies, users) and when (this is being written in June 2005) you ask. Moreover, answers might not depict a wholly homogeneous view, even within parties belonging to the same group. Fortunately, when trying to define 4G, one can find a number of similar visions that are shared by several of the forces behind 4G. This encourages us to think that a global common understanding on 4G can be attained. To show both the diversity and commonality of viewpoints on 4G, a few answers to the above question are shown.

1. Unlike 3G, which refers to a specific mobile standard and allows the transfer of data at a minimum accepted speed, 4G, or the fourth generation of wireless communications, refers to a collection of technologies and standards that will find their way into a range of new ubiquitous computing and connections systems. In its earliest stages, 4G offers the promise of allowing users to connect to the Internet and one another through a variety of devices and standards anytime, anywhere, and at a wide range of speeds, from narrowband to broadband [95, 96].

2. 4G is defined as mobile telephony at data rates of 100 Mbps globally; that is, between any two points in the world. Locally, 1 Gbps will be possible [97].

3. The main characteristics of the next generation of mobile systems are [98]:

   - A new air interface aiming for 50 to 100 Mbps deployable in 2010. (This is what defines 4G.)

   - The integration of existing system to interwork with each other and with the new interface. (This is what defines system beyond 3G.)

4. The European vision for a fourth generation terrestrial system is a fully IP-based integrated system offering any kind of services at any time and able to support multiple classes of terminals. In order to accommodate future services that require high capacity, a broadband component is envisioned with a target peak data cell throughput of more than 20 Mbps in vehicular environments, using a 50- to 100-MHz bandwidth [99].

5. 4G is defined as a network that is spectrum efficient and that can provide high speed services over the air and offer seamless interoperability, roaming, and real content. It is always on, technologically transparent, and affordable with reasonable QoS and high security (100).

6. 4G is an "ultra high-speed wireless network," an information super-highway without cables. The new network will enable wireless, three-dimensional augmented, and virtual reality connection between phone users [101].

7. 4G is "wireless ad hoc peer-to-peer networking" [102].

8. The final definition of 4G will have to include something as simple as this: if a consumer can do something at home or in the office while wired to the Internet, those consumers must be able to do it wirelessly in a fully mobile environment [102].

9. 4G is a system of systems that can take advantage of all kinds of different wireless technology [101].

10. Some would call 4G mobile broadband services, but that was supposedly the definition of 3G. An alternate proposition is that 4G represents the multinetwork environment of WWAN and WLAN [103].

11. The definition of 4G that ITU-R approved in June 2003 states that the data rates should be around 100 Mbps when moving fast (like in a train) and 1 Gbps when not moving [104].

Based on the above 4G characterizations and several others [105-109], one may attempt to summarize the visions into a single definition, encompassing key features foreseen for 4G. The 4G will be a fully IP-based integrated *system of systems* and *network of networks* achieved after the convergence of wired and wireless networks as well as computer, consumer electronics, communication technology, and several other convergences that will be capable of providing 100 Mbps and 1 Gbps, respectively, in outdoor and indoor environments with end-to-end QoS and high security, offering any kind of services anytime, anywhere, at affordable cost and one billing.

1.4 Scientific Approach

In order to develop an Adaptive 4G Global-Net (A4GN) (see Figure 1.10), the following six key elements or enabling technologies have been identified:

1. Adaptive and scalable air interfaces;

2. Reconfigurable ambient networks;

3. End-to-end security and QoS;

4. Highly available backbone technologies (e.g., Fiber Ring, MPLS);

5. User friendly multimedia interfaces and context-aware technologies;

6. Flexible platforms.

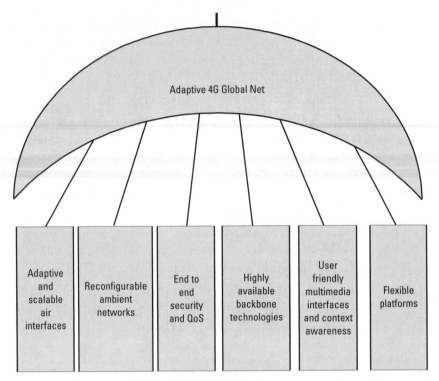

Figure 1.10 Research umbrella.

For each of the highlighted issues, a more detailed description of the research to be done is provided in the following sections.

1.4.1 Adaptive and Scalable Air Interfaces

In order to justify the need for a new air interface, target requirements should be set high enough to ensure that the system will be able to serve long into the future. Considering the emerging technologies and the types of services that will need to be provided in the future communication system, a reasonable goal would be to aim for 100-Mbps full mobility wide area coverage and 1 Gbps for low mobility indoor coverage. Currently, the provision of highly reliable wireless link technologies envisages the use of access technologies (e.g., orthogonal frequency division multiplexing (OFDM) and multi-carrier code division multiple access (MC-CDMA), multiuser modulation, efficient coding schemes, reconfigurable radio, multiple antennas, and adaptive power control. To enable adaptability to network conditions and scalability (in terms of data rate and quality of service), new air interfaces (e.g., 3G+ like high speed downlink packet access, multicarrier and multitone CDMA, ultra-wideband, and optical wireless systems) have to be investigated in terms of coexistence with other radio systems, multiple access capability, resistance to interferers, low cost, low power consumption, and implementation issues. The research-related areas are shown in Figure 1.11.

1.4.2 Reconfigurable Ambient Networks

To enable the advent of ambient intelligence, the underlying ambient networks need a high degree of reconfigurability. Reconfigurable networks imply the design of new communications mechanisms at different layers. These include research in the areas shown in Figure 1.12.

The reconfiguration actions have an impact on various levels of mobile systems architecture and introduce high complexity that has to be handled by some reconfigurability management intelligence (distributed or not). A possible approach is to start by describing evolutionary scenarios based on existing systems and gradually propose new scenarios deploying leading-edge technologies. Research should start with the definition of user requirements, regulatory demands, system methodology, and constraints. Since reconfigurability concepts must be compatible to legacy networks and also be embedded into future IP-based networks and future network topologies (such as ad hoc networks), the support by reconfigurable routers must be ensured.

Results derived from reconfigurable ambient networks and research on intelligent packet transfer techniques may be applied. This may help to identify new higher layer (i.e., above the physical layer) concepts and algorithms to cope

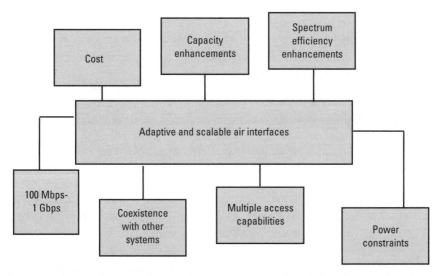

Figure 1.11 Research areas related to the development and validation of new air interfaces.

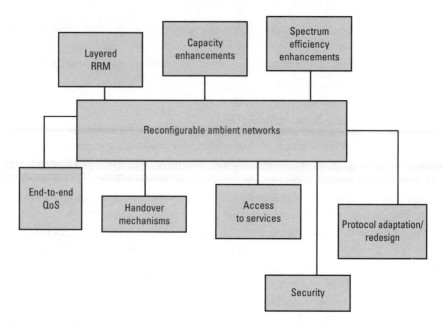

Figure 1.12 Research areas related to reconfigurable networks.

in an efficient manner with frequent topology variations, end user requirements, resource availability, and power constraint considerations. To this end, the development and optimization of the above layered and distributed resource management and radio link control algorithms, along with the appropriate middleware, can be conceived. Additionally, methods for spectrum sharing, mode detection, seamless service provision by mode switching, and secure software download in a combination of classical methods, such as mobile radio design and computer science (a very broad term), may be developed. This creates a synergy platform for novel concepts, which can be combined with a full consideration of the real user needs.

1.4.3   Security Across All Layers

To enable end-to-end security in A4GN, the system will be characterized in terms of:

- *Availability*, which ensures the survivability of network services despite denial of service attack. *Authentication*, which enables a node to ensure the identity of the peer node that it is communicating with. *Confidentiality*, which ensures that certain information will be never disclosed to unauthorized entities, without proper authentication.

- *Integrity*, which guarantees that a message is the exact replica of the original source when it is delivered to the destination.

- *Authorization*, which is the process of deciding if device X is allowed to have access to service Y. It must be mentioned that devices that form ad hoc networks are more exposed to phising, spoofing that leads to failure, and require rugged security measures.

- *Privacy*, which can prevent the information of the users from flowing to others.

- *Flexibility*, which allows different upper protocols or applications to enforce their own security policy

Furthermore, novel cryptography algorithms suitable for wireless communication and low energy implementation should be developed.

1.4.4 Highly Available Backbone Technologies

Wireless systems have to be interconnected using a long-haul backbone network that is IP-based in order to provide global coverage. The network concepts are moving from network-centred towards person-centred solutions, introducing the new network paradigm of WPANs. The idea is illustrated in Figure 1.13.

The main concept is represented by the short-range network solution where the person within his or her personal space can establish connectivity to his or her personal devices and the outer world. The backbone network will be based on optical technology, complemented by a collection of other wired and wireless systems. In this sense, mechanisms for the provision of interoperability

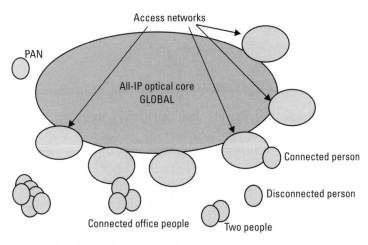

Figure 1.13 The future Internet.

between existing and forthcoming wireless infrastructures, with the optical mechanism supporting the IP flows, become mandatory.

Research in this area includes that of (1) mobility mechanisms (to ensure successful interaction among WPANs, the appropriate mechanism must be guaranteed for a dynamic type of communication), and (2) the coexistence of WPANs and other wireless devices based on various standards. This issue should be addressed by evaluating quantitatively the effects of the coexistence model consisting of the following four sections: physical layer models, medium access control (MAC) layer models, radio frequency (RF) channel models, and data traffic models.

An initial approach aims at understanding how different wireless services operating in the same band may affect each other. Later, coexistence mechanisms are established. Depending on the operating environment, one mechanism may be preferred over the other.

1.4.5   User Friendly Multimedia Interfaces and Context-Aware Technologies

Interfaces like smart cards with biometric capabilities will play an important role in context-aware scenarios. The profile of the user (including not just the network parameters but also favorite URLs, e-mail address book, customer bank profile, and so forth) will be saved in smart cards through which the user will transparently interact with the communications infrastructures around him with added security. In parallel, this interface will be used by operators as a billing support and for offering value-added services, depending on location parameters and profile. Additionally, other ways of interaction based on speech and/or gestures, which are defined as multimodal interactive techniques, need to be planned for these scenarios. More research topics for speech, spoken language, and multimedia technologies exist in the context of deployment of hand-held devices that will be used for short or long-range communications.

1.4.6   Flexible Platforms

In order to cope with the broad range of end user requirements in a context-aware environment, a flexible architecture has to be designed. Flexible architecture means that, to accommodate a wide spectrum of devices and functionalities in wireless systems, the underlying hardware platforms need a high degree of reconfigurability, including not only field programmable gate array (FPGA) technology, but also mixed architectures based on parallel processing, digital signal processing (DSP), ASICs, mixed signal integrated circuit (IC), and hardware IP blocks. An appropriate design flow (including all algorithmic stages up to the system level) should enable the codesign of both various specific devices and flexible high-end multisystem devices. This design flow, to ensure flexibility and taking into account various constraints (low power, low cost, low form factor), relies heavily on embedded systems and system-on-a-chip concepts. Last but not least, the introduction of multisystem devices is only possible with the help of a high degree of reconfigurable antennas and RF subsystems, as well as of research on new RF architectures (e.g., zero-IF). The different areas of relevance are shown in Figure 1.14.

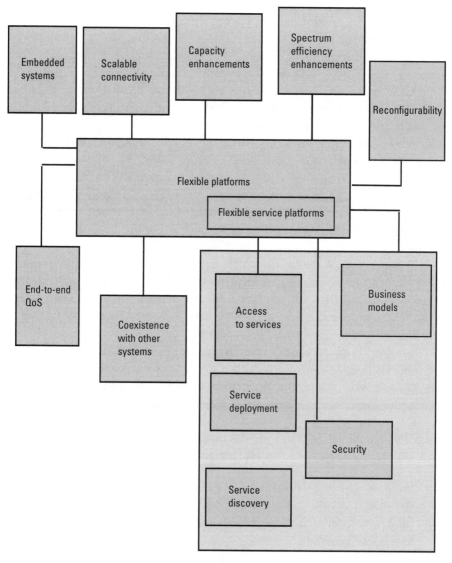

Figure 1.14 Development of flexible platforms.

1.5   Preview of the Book

The book consists of seven chapters, concentrating mainly on 4G wireless communication systems. It goes deeper into the technical difficulties of 4G implementation as well as challenges to achieving 4G. Chapter 2 approaches 4G from a user-centric perspective. This chapter introduces and discusses user needs and trends in order to get a better insight into what user expectations are for future wireless communication systems. Specific and detailed discussions on 4G scenarios and services are also presented. ITU visions as well as our own vision are presented and discussed. Finally spectrum issues are introduced. The first and foremost challenge has always been to select a suitable access technique for defining and developing any mobile communications generation. The key access techniques based on multicarrier techniques are presented in Chapter 3.

Chapter 4 introduces the key technologies for 4G: MIMO technology, Radio Resource Management, Software Radio Communication System, Mobile IP, and Relaying Techniques.

Major research initiatives focusing on 4G are presented in Chapter 5. These are the Wireless World Research Forum (WWRF), Mobile IT Forum (mITF), Future Technology Universal Radio Environment (FuTURE) project, Next Generation Mobile Communication Forum (NGMC), 4G Research Cooperation Projects in the European Sixth Framework Programme, Worldwide Wireless Initiative (WWI), Samsung 4G Forum, and the eMobility Technology Platform.

Chapter 6 explores 4G terminals from the technology and application point of view, in an attempt to identify possible trends, promising technical solutions and challenges for designers of future portable communications equipment. The critical success factors for 4G terminals are service convergence, a user-centric interface, and portable intelligence.

Finally, Chapter 7 introduces a fascinating vision of 4G that appears to be the most viable solution for achieving a true knowledge society, as well as for serving users, while at the same time, being attractive to industry, operators, and service providers.

References

[1]   Prasad, R., *OFDM for Wireless Communication Systems*, Norwood, MA: Artech House, 2004.

[2]   Prasad, R., and M. Ruggieri, *Technology Trends in Wireless Communications*, Norwood, MA: Artech House, 2003.

[3]   Pereira, J. M., "Balancing Public and Private in Fourth Generation," Proceed. IEEE PIMRC 2001, San Diego, September/October 2001, pp. 125-132.

[4]   ETSI (1991b), *General Description of a GSM PLMN*, European Telecommunications Standards Institute, GSM recommendations 01.02.

[5]   Lin, Y.-B., and I. Chlamtac, *Mobile Network Protocols and Services*, New York: Wiley, 2000.

[6]   Krenik, W. R., "Wireless User Perspectives in the United States," Wireless Personal Communications Journal, Kluwer, Vol. 22, No. 2, August 2002, p.p. 153–160.

[7]   Schiller, J., *Mobile Communications*, Reading, MA: Addison-Wesley, 2000.

[8]   Prasad, N. R., "GSM Evolution Towards Third Generation UMTS/IMT-2000," Proc. IEEE International Conference on Personal Wireless Communication, 1999, pp. 50-54.

[9]   Nakajima, N., "Future Communications Systems in Japan," *Wireless Personal Communications*, Vol. 17, No. 2, June 2001, pp. 209-223.

[10] Digital Cellular Telecommunications System (Phase 2+); General Packet Radio Service (GPRS); Overall Description of the GPRS Radio Interface; Stage 2. GSM 03.64, Version 7.0.0, Release 1998.

[11] Cai, J., and D. J. Goodman, "General Packet Radio Service in GSM," *IEEE Communications Magazine*, October 1997.

[12] Lin, P., and Y. Lin, "Channel Allocation for GPRS," *IEEE Transaction on Vehicular Technology*, Vol. 50, No. 2, March 2001, pp. 375-387.

[13] Priggouris, G., S. Hdjiefthymiades, and L. Merakos, "Supporting IP QoS in the General Packet Radio Service," *IEEE Network*, September/October 2000, pp. 8-17.

[14] Sarikaya, B., "Packet Mode in Wireless Networks: Overview of Transition to Third Generation," *IEEE Communication Magazine*, Vol. 38, No. 9, September 2000, pp. 164-172.

[15] Muratore, F., *UMTS: Mobile Communications for Future*, New York: John Wiley & Sons, 2000.

[16] Lin, Y. B., A. Pang, and M. F. Chang, "VGPRS a Mechanism for Voice over GPRS," Proc. International Conference on Distributed Computing Systems Workshop, 2001, pp. 435-440.

[17] 3rd Generation Partnership Project (3GPP), "UMTS and GSM Standards Data Base," available at http://www.3gpp.org.

[18] van Nobelen, R., et al., "An Adaptive Radio Link Protocol with Enhanced Data Rates for GSM Evolution," *IEEE Personal Communications*, Vol. 6, No. 1, February 1999, pp. 54-64.

[19] Furuskar, A., et al., "EDGE: Enhanced Data Rate for GSM and TDMA/136 Evolution," *IEEE Personal Communications*, Vol. 6, No. 3, June 1999, pp. 56-66.

[20] Berthet, A., R. Visoz, and P. Tortelier, "Sub-Optimal Turbo-Detection for Coded 8-PSK Signals over ISI Channels with Application to EDGE Advanced Mobile System," Proc. 11th IEEE International Symposium on Personal, Indoor and Mobile Radio Communications, PIMRC 2000, Vol. 1, 2000, pp. 151-157.

[21] Gerstacker, W. H., and R. Schober, "Equalisation for EDGE Mobile Communications," *Electronics Letters*, Vol. 36, No. 2, January 2000, pp. 189-191.

[22]   Prasad, R., *Wideband CDMA for Third-Generation Mobile Systems*, Norwood, MA: Artech House, 1998.

[23]   Prasad, R., W. Mohr, and W. Konhauser, *Third Generation Mobile Communication System*, Norwood, MA: Artech House, 2000.

[24]   Ojanpera, T., and R. Prasad, *WCDMA: Towards IP Mobility and Mobile Internet*, Norwood, MA: Artech House, 2001.

[25]   Prasad, R., Towards a Global 3G System-Advanced Mobile Communications in Europe, Norwood, MA: Artech House, 2001.

[26]   Schiller, J., *Mobile Communications*, Reading, MA: Addison-Wesley, 2000.

[27]   Prasad, N. R., and A. Prasad, *WLAN Systems and Wireless IP for Next Generation Communication*, Norwood, MA: Artech House, 2002.

[28]   Rapeli, J., "Future Directions for Mobile Communications Business, Technology and Research," *Wireless Personal Communications*, Vol. 17, No. 2, June 2001, pp. 155-173.

[29]   Zeng, M., A. Annamalai, and V. Barghava, "Recent Advances in Cellular Wireless Communications," *IEEE Communications Magazine*, September 1999, Vol. 37, No. 9, pp. 128-138.

[30]   Holma, H., and A. Toskala, *WCDMA for UMTS*, New York: John Wiley & Sons, 2000.

[31]   Dinan, E. H., and B. Jabbari, "Spreading Codes for DS-CDMA and Wideband CDMA Cellular Networks," IEEE Communications Magazine, September 1998, pp. 48-54.

[32]   Kauffmann, P., "Fast Power Control for Third Generation DS-CDMA Mobile Radio System," *Proc. 2000 International Zurich Seminar on Broadband Communications*, 2000, pp. 9-13.

[33]   Jorguseski, L., J. Farserotu, and R. Prasad, "Radio Resource Allocation in Third Generation Mobile Communication Systems," *IEEE Communication Magazine*, Vol. 39, No. 2, February 2001, pp. 117-123.

[34]   Benedetto, S., and G. Montorsi, "Unveiling Turbo Codes: Some Results on Parallel Concatenated Coding Schemes," *IEEE Transactions on Information Theory*, Vol. 42, No. 2, 1996, pp. 409-428.

[35]   Perez, L. C., J. Seghers, and D. J. Costello, Jr., "A Distance Spectrum Interpretation of Turbo Codes," *IEEE Transactions on Information Theory*, Vol. 42, 1996, pp. 1698-1709.

[36]   3rd GPP, Technical Specification Group, Radio Access Network, Working Group 1, Multiplexing and Channel Coding (FDD).

[37]   Knisely, D. N., et al., "Evolution of Wireless Data Services: IS-95 to cdma2000," *IEEE Communications Magazine*, Vol. 36, No. 10, October 1998, pp. 140-149.

[38]   Rao, Y. S., and A. Kripalani, "cdma2000 Mobile Radio Access for IMT 2000," IEEE International Conference on Personal Wireless Communication, 1999, pp. 6-15.

[39]   Lee, D., H. Lee, and L. B. Milstein, "Direct Sequence Spread Spectrum Walsh-QPSK Modulation," *IEEE Transaction on Communications*, Vol. 46, No. 9, September 1998, pp. 1227-1232.

[40] Chulajata, T., and H. M. Kwon, "Combinations of Power Controls for cdma2000 Wireless Communications System," *Proc. 52nd Vehicular Technology Conference*, 2000, IEEE VTS Fall VTC2000, 2000, pp. 638-645.

[41] C.-L. I., and S. Nanda, "Load and Interference Based Demand Assignment for Wireless CDMA Networks," Proc. IEEE Globecom, 1996.

[42] "Standards for cdma2000 Spread Spectrum Systems," EIA/TIA IS-2000, pp. 1-6.

[43] Etemad, K., "Enhanced Random Access and Reservation Scheme in CDMA2000," *IEEE Personal Communications*, April 2001, pp. 30-36.

[44] IEEE 802.11, "IEEE Standard for Wireless LAN Medium Access Control (MAC) and Physical Layer (PHY) Specifications," November 1997.

[45] Crow, B. P., et al., "IEEE 802.11 Wireless Local Area Network," *IEEE Communications Magazine*, September 1997, pp. 116-126.

[46] ETSI, "Radio Equipment and Systems, High Performance Radio Local Area Network (HIPERLAN) Type 1," European Telecommunication Standard, ETS, 300-652, October 1996.

[47] van Nee, R., and R. Prasad, *OFDM for Wireless Multimedia Communications*, Norwood, MA: Artech House, 2000.

[48] Macker, J. P., and M. S. Corson, "Mobile Ad Hoc Networking and the IETF," Proc. ACM, Mobile Computing and Communications, Vol. 2, No. 1, January 1998.

[49] Xu, S., and T. Saadawi, "Does the IEEE 802.11 MAC Protocol Work Well in Multihop Wireless Ad Hoc Network?" *IEEE Communications Magazine*, Vol. 39, No. 6, June 2001, pp. 130-137.

[50] Niemegeers, I. G., and S. M. H. de Groot, "From Personal Area Networks to Personal Networks: A User Oriented Approach," *Wireless Personal Communications*, Vol. 22, Issue 2, August 2002, pp. 175–186.

[51] Bisdikian, C., "An Overview of the Bluetooth Wireless Technology," *IEEE Communications Magazine*, Vol. 39, No. 12, December 2001, pp. 86-94.

[52] "Bluetooth 2000: To Enable the Star Trek Generation," Cahners In-Stat Group, MM00-09BW, June 2000.

[53] Bluetooth Special Interest Group, "Specification of the Bluetooth System," December 1999.

[54] Siep, T. M., et al., "Paving the Way for Personal Area Network Standards: An Overview of the IEEE P802.15 Working Group for Wireless Personal Area Networks," *IEEE Personal Communications*, Vol. 7, No.1, February 2000, pp. 37 -43.

[55] Jha, U., "Wireless Landscape-Need for Seamless Connectivity," *Wireless Personal Communications*, August 2002, pp. 275–283.

[56] Japanese Telecommunications Technology Council, "A Partial Report on Technical Conditions on Next-Generation Mobile Communications System," September 1999.

[57]   Ohmori, S., Y. Yamao, and N. Nakajima, "The Future Generations of Mobile Communications Based on Broadband Access Methods," *Wireless Personal Communications*, Vol. 17, No. 2, June 2001, pp. 175-190.

[58]   Mohr, W., "Development of Mobile Communications Systems Beyond Third Generation," *Wireless Personal Communications*, Vol. 17, No. 2, June 2001, pp. 191-207.

[59]   Chaudhury, P., W. Mohr, and S. Onoe, "The 3GPP Proposal for IMT-2000," *IEEE Communications Magazine*, Vol. 37, No. 12, 1999, pp. 72-81.

[60]   Poor, H. V., and G. W. Wornell, *Wireless Communications-Signal Processing Perspective*, Englewood Cliffs, NH: Prentice Hall, 1998.

[61]   Andersen, J. B., "Array Gain and Capacity for Known Random Channels with Multiple Element Arrays at Both Ends," *IEEE Journal on Selected Areas in Communications*, Vol. 18, No. 11, November 2000, pp. 2172-2178.

[62]   Foschini, G. J., and M. J. Gans, "Capacity when Using Multiple Antennas at Transmit and Receive Sites and Rayleigh-Faded Matrix Channel Is Unknown to the Transmitter," in *Advanced in Wireless Communications*, J. M. Holtzmann and M. Zorzi (Eds.), Boston, MA: Kluwer Academic Publishers, 1998.

[63]   IEEE Personal Communications, Vol. 5, February 1998.

[64]   Pattan, B., *Robust Modulation Methods and Smart Antennas in Wireless Communications*, Englewood Cliffs, NJ: Prentice Hall, 2000.

[65]   Wolniansky, P. V., et al., "V-BLAST: An Architecture for Realizing Very High Data Rates over the Rich-Scattering Wireless Channel," *Proc. ISSSE-98*, Pisa, Italy, September 29, 1998.

[66]   Foschini, G. J., "Layered Space-Time Architecture for Wireless Communication in a Fading Environment when Using Multi-Element Antennas," *Bell Labs Technical Journal*, 1996, pp. 41-59.

[67]   Sheikh, K., et al., "Smart Antennas for Broadband Wireless Access Network," *IEEE Communications Magazine*, Vol. 37, No. 11, 1997, pp. 100-105.

[68]   Andersen, J. B., "Role of Antennas and Propagation for the Wireless System Beyond 2000," *Wireless Personal Communications*, Vol. 17, No. 2-3, 2001, pp. 303-310.

[69]   Tarokh, V., N. Seshandri, and R. C. Calderbank, "Space-Time Codes for High Data Rate Wireless Communication: Performance Criteria and Code Construction," *IEEE Transaction on Information Theory*, Vol. 44, No.2, 1998, pp. 744-765.

[70]   Nanda, S., K. Balachandran, and S. Kumar, "Adaptation Techniques in Wireless Packet Data Services," *IEEE Communications Magazine*, Vol. 38, No. 1, 2000, pp. 54-64.

[71]   Berrou, C., A. Glavieux, and P. Thitimajshima, "Near Shannon Limit Error-Correcting Coding and Decoding: Turbo-Codes," Proc. ICC93, May 1993.

[72]   Benedetto S., and G. Montorsi, "Unveiling Turbo Codes: Some Results on Parallel Concatenated Coding Schemes," *IEEE Transactions on Information Theory*, Vol. 42, No. 2, March 1996, pp. 409-428.

[73] Robertson, P., and T. Worz, "A Novel Bandwidth Efficient Coding Scheme Employing Turbo Codes," *Proc. ICC96,* June 1996.

[74] Benedetto, S., et al., "Parallel Concatenated Trellis Coded Modulation," *Proc. ICC96,* June 1996.

[75] Benedetto, S., et al., "Serial Concatenated Trellis Coded Modulation with Iterative Decoding: Design and Performance," *Proc. Comm. Theory Miniconf. 97,* November 1997.

[76] Mitchell, T., "Broad Is the Way [Ultra-Wideband Technology]," *IEE Review,* Vol. 47, No. 1, January 2001, pp. 35-39.

[77] Verdù S., *Multiuser Detection,* Cambridge University Press: Cambridge, U.K., 1998.

[78] Buzzi, S., M. Lops, and A. M. Tulino, "MMSE Multi-User Detection for Asynchronous Dual Rate Direct Sequence CDMA Communications," *Proc. 9th PIMRC,* Boston, MA, September 1998.

[79] Mitra, U., "Comparison of Maximum Likelihood-Based Detection for two Multi-Rate Access Schemes for CDMA Signals," *IEEE Transactions on Communications,* Vol. 47, January 1999, pp. 64-77.

[80] Buzzi, S., M. Lops, and A. M. Tulino, "Blind Adaptive MMSE Detection for Asynchronous Dual-Rate CDMA Systems: Time-Varying Versus Time-Invariant Receivers," *Proc. IEEE GLOBECOM'99,* December 1999.

[81] Bensley, S. E., and B. Aazhang, "Subspace-Based Channel Estimation for Code-Division Multiple-Access Communication Systems," *IEEE Transactions on Communications,* Vol. 44, August 1996, pp. 1009-1020.

[82] Weiss, A. J., and B. Friedlander, "Channel Estimation for DS/CDMA Downlink with Aperiodic Spreading Codes," *IEEE Transactions on Communications,* Vol. 47, October 1999, pp. 1561-1570.

[83] Ramakrishna S., and J. M. Holtzman, "A Scheme for Throughput Maximization in a Dual-Class CDMA System," *IEEE Journal on Selected Areas on Communications,* Vol. 16, August 1998, pp. 830-844.

[84] Dziong Z., M. Jia, and P. Mermelstein, "Adaptive Traffic Admission for Integrated Services in CDMA Wireless-Access Networks," *IEEE Journal on Selected Areas on Communications,* Vol. 14, December 1996, pp. 1737-1747.

[85] Soroushnejad, M., and E. Geraniotis, "Multi-Access Strategies for an Integrated Voice/Data CDMA Packet Radio Network," *IEEE Transactions on Communications,* Vol. 43, February/March/April 1995, pp. 934-945.

[86] Caceres, R., and V. N. Padmanabhan, "Fast and Scalable Handoffs for Wireless Internetworks," *Proc. of ACM MobiCom'96,* November 1996.

[87] Benvenuto, N., and F. Santucci, "A Least Squares Path Loss Estimation Approach to Handover Algorithms," *IEEE Transactions on Vehicular Technology,* Vol. 48, March 1999.

[88] Graziosi, F., and F. Santucci, "Analysis of a Handover Algorithm for Packet Mobile Communications," *Proc. ICUPC'98,* Florence, Italy, October 1998, pp. 769-774.

[89]   Efthymiou, N., Y. F. Hu, and R. E. Sheriff, "Performance of Intersegment Handover Protocols in a Integrated Space/Terrestrial-UMTS Environment," *IEEE Transactions on Vehicular Technology*, Vol. 47, No. 4, November 1998, pp. 1179-1199.

[90]   Maral, G., et al., "Performance Analysis of a Guaranteed Handover Service in a LEO Constellation with a Satellite-Fixed Cell System," *IEEE Transactions on Vehicular Technology*, Vol. 47, No. 4, November 1998, pp. 1200-1213.

[91]   Lilleberg, J., and R. Prasad, "Research Challenging for 3G and Paving the Way for Emerging New Generation," *Wireless Personal Communications*, Vol. 17, No. 2, June 2001, pp. 355-362.

[92]   Farserotu, J., et al., "Scalable, Hybrid Optical-RF Wireless Communication System for Broadband and Multimedia Service to Fixed and Mobile Users," Wireless Personal Communications, Boston, MA: Kluwer Academic Publishers, Vol. 24, Issue 2, 2003, pp. 327–329.

[93]   Lauridsen, O. M., and A. R. Prasad, "User Needs for Services in UMTS," Invited Paper, International Journal on Wireless Personal Communications, Boston, MA: Kluwer Academic Publishers, Vol. 22, No. 2, August 2002, pp. 187-197.

[94]   Niemegeers, I. G., and S. M. Heemstra de Groot. "Research Issues in Ad-Hoc Distributed Personal Networks," International Journal on Wireless Personal Communications, Boston, MA: Kluwer Academic Publishers, September 2003, Vol. 26, No. 2–3, pp.149–167.

[95]   Pyramid Research Consulting, "North Asia's 4G Frontier: The Coming of Age for Asia's Telecom Industry," available at http://www.pyramidresearch.com/, June 2003.

[96]   Niemegeers, I. G., and S. M. Heemstra de Groot, "4G and Ubiquitous Computing and Communication," Sixth Strategic Workshop, Convergence Towards 4G, Rome, Italy, June 17-19, 2004.

[97]   Nielsen, T. T., and R. H. Jacobsen, "IP Opportunities in Communication Beyond 3G," Sixth Strategic Workshop, Convergence Towards 4G, Rome, Italy, June 17-19, 2004.

[98]   Schoo, P., "Mobile Adventure: IT Security in Next Generation of Mobile Communication System," Sixth Strategic Workshop, Convergence Towards 4G, Rome, Italy, June 17-19, 2004.

[99]   Kaiser, S., et. al., "4GMG-CDMA Multi Antenna Systems on Ship for Radio Enhancement (4MORE)," Proceedings IST Mobile & Wireless Communication Summit 2004, Lyon, France, June 27-30, 2004, pp. 1069-1073.

[100]   Ravikumar, J., "Challenges in Deploying 3G Networks and Beyond," CoE Conference, Nanyang Technological University, Singapore, March 2003.

[101]   Bartlett, H., "The Evolving Network," available at http://www.thefeature.com.

[102]   Kupetz, A. H., and K. T. Brown, "4G-A Look into the Future of Wireless Communications," Rollings Business Journal, 2003.

[103]   Cassidy, S., "3G's Feeling the Heat," Loop Internet magazine, 2003.

[104]   Nozawa, T., "NTT DoCoMo's 4G Test Results in 300Mbps Data Rate in Moving Car," NE ASIA online, June 1, 2004.

[105] Proceedings IST Mobile & Wireless Communication Summit 2004, Lyon, France, June 27-30, 2004.

[106] "Toward Open Wireless Architecture," special issue on 4G Mobile Communication, IEEE Wireless Communications, Vol. 11, No. 2, April 2004.

[107] Financial Times, June 8, 2004.

[108] Budder, R., "Getting Access Lines: New Technology Paves the Way for Convergence Between Fixed and Mobile Telecoms," Financial Times, June 8, 2004, p. 11.

[109] Lilleberg, J., and R. Prasad, "Research Challenges for 3G and Paving the Way for Emerging New Generations," Wireless Personal Communication, an International Journal, Kluwer, Vol. 17, June 2001, pp. 355-362.

[110] Kim, Y., et al., "Beyond 3G: Vision, Requirements, and Enabling Technologies," IEEE Communications Magazine, Vol. 41, No. 3, March 2003, pp. 120-124.

2

A User-Centric Approach to 4G: Visions and Foresights

2.1 Introduction and Motivation

2.1.1 Introduction

After briefly introducing some basic concepts of 4G in Chapter 1, this chapter explores 4G in more detail, focusing on 4G visions and foresights, as well as on the shape that 4G could take by means of scenario-based methodology. Other important factors like user needs and trends as well as spectrum identifications for 4G will also be discussed. A user-centric approach to 4G is assumed in this chapter.

The deployment of International Mobile Telecommunications-2000 (IMT-2000) networks has started in several countries, and high-speed multimedia services are currently being provided and further developed. However, even at this introductory stage of 3G systems, enhancement activities have already started on a global scale. In fact, for several years already there have been active discussions and exchange of visions about systems beyond IMT-2000, commonly referred to as the fourth generation, or 4G for short, in various forums and organizations as well as in the open literature. The reasons for this early interest abound. One of the main driving forces for such a development is, without a doubt, the unrelenting demand for higher data rates and ubiquitous wireless access that the planned and foreseen applications and services are expected to generate. Equipment manufactures are in a key position and need to react promptly by developing technically sound and appealing solutions, capable of fulfilling users' and operators' expectations, within a common and global

29

framework. This is a very complex procedure, where key manufacturing players must find a good balance between achieving a mutually convenient technical consensus, interpreting market needs, and imposing industry views. More than ever, academia is greatly contributing by identifying and investigating promising component technologies for future wireless communication systems. Operators and service developers have a huge stake in offering advanced and appealing wireless communication services to their customers, while aiming at profitable operations. Regulators, also at the front of the 4G development scene, control, normalize, and create policies and prosperous environments aiming towards the development of a widely accepted framework. Finally, conveying the voices of all involved parties, prestandardization bodies and discussion forums have a key role in finding a global consensus, as they serve not only as the melting pots of visions but also as the forefront stage for technical debate. Considering the interaction among the aforementioned key players and taking into account that these diverse contributors do not necessarily share the same interests, goals, and time plans, no one would be surprised to realize that finding a universal definition of 4G is a very elusive task, even after several years of activities and countless attempts in the literature.

2.1.2 Motivation for the Development of a User-Centric View

The number of mobile subscribers worldwide has increased from more than 200 million in 1997 to nearly 950 million in 2001 and 1.8 billion in 2005. Figure 2.1 shows the current global growth of mobile subscribers with some predictions for the near future. Mobile phones are no longer regarded as expensive communication devices only for a minority of users that can afford them. Nowadays, the mobile phone has become an object of necessity and convenience widely used by people of all conditions and age.

Several reasons could account for such an impressive growth. Competition among manufacturers has resulted in substantial price drops in terminals, with affordable models available to all segments of the market. These factors have significantly contributed to an explosive growth in traffic. On the other hand, people using wireline broadband Internet are likely to expect similar degrees of data connectivity on the move, with comparable data speeds and quality of service. It could be anticipated that if user expectations for new, useful, and exciting mobile services and applications are matched by the industry with capable and appealing terminals and supporting networks, we can expect a future with even higher numbers of subscribers as well as a steep increase in the amount of data traffic. It is widely accepted that users will expect a more and more dynamic, continuing stream of new applications, capabilities, and services in the future. These services and applications will exploit the fact that terminals will support high data rates, and they will also fully take advantage of other capabilities, like

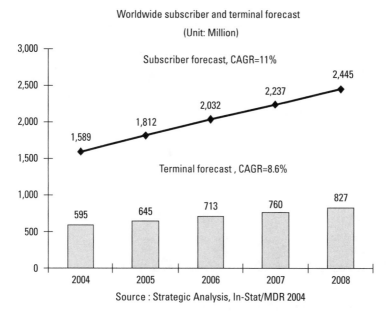

Worldwide subscriber and terminal forecast

(Unit: Million)

Source : Strategic Analysis, In-Stat/MDR 2004

Figure 2.1 The global growth of mobile subscribers and terminals. (CARG: compound annual growth rate.)

seamless connectivity to different access networks or peer terminals, multiple air interfaces on board allowing simultaneous access, positioning, higher processing power and larger memory capacity of terminals, and advanced imaging.

Rather than provide a precise definition of the future 4G wireless communication systems, this chapter attempts to describe an updated overview of the 4G landscape of visions and foresights. Indeed, at this time there is no complete, unique, and widely accepted definition of 4G, but rather a large collection of prospective descriptions, each possibly biased toward the business interests of the defining party. In order to predict properly the 4G era, an objective methodology is necessary. A scenario-based methodology will be introduced for such a purpose. The present chapter discusses some of the key 4G visions as well as other tightly related factors, chiefly spectrum issues and their close interaction with 4G developments. As an introductory illustration, Figure 2.2 depicts the principal driving forces behind past, current, and future wireless communication systems, with an emphasis on cellular evolution.

This chapter is organized as follows. Section 2.2 introduces and discusses user needs and trends in order to get a better insight into what user expectations on 4G might be. The key driving force justifying or creating the need for developing 4G is *service*. We will shed some light on the area of future services by considering first a number of likely scenarios for the future, all having as a starting point the needs (and trends) of the users. Section 2.3 introduces us to

Key drivers (requirements)

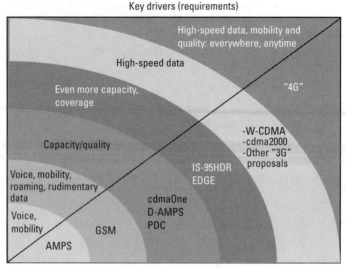

Technology response

Figure 2.2 Wireless market drivers and technology responses.

scenario and service development, while more specific and detailed discussions on 4G scenarios and services for the identified scenarios are discussed in Sections 2.4 and 2.5, respectively. Visions and foresights of 4G are discussed in Section 2.6 These include discussions on ITU visions and other approaches trying to define 4G. Finally, spectrum issues are discussed in Section 2.7.

2.2   User Needs and Trends

This section emphasizes the importance of the service-driven approach to 4G and briefly discusses user needs and trends from the views of user, market, and technology.

2.2.1   A Service-Driven Approach to 4G

Mobile communication systems have steadily evolved in a relatively short period, spanning mainly the last 25 years. Today, it is widely accepted that services will be the most important success factor for the telecommunication businesses to come. Therefore, it is anticipated that next generation communication systems will be service-driven and user-centric-service development drives technological development, not the other way around.

Until now, wireless communication systems were developed around a given technology and then the appropriate services were developed for these systems. With this *bottom-up approach*, commercialization success cannot be guaranteed as the services offered match the technical capabilities of the systems but do not necessarily match the expectation of the users. This approach has clearly worked in the second generation (2G) where the basic services were broadly introduced and the level of expectation was nearly fulfilled with voice and text messaging services. The same cannot necessarily be said with the 3G, where the gap between services and user expectations has been difficult to close, since users have been exposed to sophisticated services through wired high-speed Internet connections. As 4G development is taking off, the wireless world is now focusing on an opposite method, a *top-down approach*, where technology follows services and applications. In order to meet user expectations and hence make of 4G a great technical and commercial achievement, it is paramount to predict what these expectations actually are. Undoubtedly, user behavior and future expectations are very difficult to predict. One cannot rely on linear extrapolations based on past patterns, making also the top-down approach somewhat risky. Service development is inherently a formidable task since the services are based upon the user needs. It is commonly agreed that 4G should be developed around the user and that technology is just a means, not the goal of the 4G development. Recently, some authors have approached 4G from such a user-centric perspective [1, 2, 11–14]. The key features that users expect from future 4G systems are obtained from a number of relevant user scenarios (e.g., business on-the-move, smart shopping, mobile tourist guides [2]). In Sections 2.3 and 2.4, a systematic framework for developing scenarios and extracting promising services from these scenarios will be presented. As a basis for developing a systematic framework, future trends need to be extracted first to cater to these user needs. In what follows, a sketch of such user, market, and technology trends is given.

2.2.2 User Needs and Trends

Users will expect new dynamic and flexible applications and services that are ubiquitous and available across a range of devices, possibly using a single subscription (or a single identity, number, or address) to access any network. Adaptable communication systems offering customized and ubiquitous services based on diverse individual needs will require flexibility in the technology in order to satisfy multiple demands simultaneously. To satisfy users, work has already begun on the convergence of telecommunication services such as digital broadcasting and commercial wireless services. Higher data rates will enable new applications and services. High speed capability is important when seen as a supporting part of the development, but this capacity is meaningless if there are no services or applications making full use of it. It is thought that most of the users are not automatically appealed to by the speed at which a wireless communication device can operate, but most fundamentally by its capacity for performing efficiently important and useful tasks (or applications). The race for developing a new technology with record-breaking data throughputs could be useful for brand marketing purposes, but it appears to be somehow short-sighted if the goal is to attract customers' attention. As seen in the Section 2.2.1, it is commonly agreed that 4G should be approached from a user-centric perspective. The procedure for extracting important 4G features likely to be appreciated by the users starts by mapping real-life situations with possible services that could help to make users' lives more comfortable and safe. Some common and useful attributes associated with these services, the sought features, are then identified. Important features valued by users are listed below:

- *User friendliness:* Users are thought to give a top priority to easy-to-use, simple, and intuitive man-machine interfaces.
- *User personalization:* Flexibility in configuring the terminal, services, and contents according to the preferences of the user is an important asset to create a sense of uniqueness in the user.
- *Network heterogeneity:* Users highly value the ubiquitous provision of connectivity and common services. In terms of future communication systems, the solution to such a requirement would be a network of networks, the 4G network, that transparently and seamlessly offers continuous and pervasive services with provision of QoS.
- *Terminal heterogeneity:* Users will enjoy a large and diverse variety of terminals, from small and simple devices with limited capabilities to high-performance multifeatured ones. The diversity in communication devices and supporting networks can be considered as one the most distinctive characteristics of 4G.

User friendliness is related to terminal design, though services and applications as well as the transparency and performance of the supporting network also play important roles in conferring satisfaction to the user. User personalization is related to the terminal, services, and applications. There is an inherent mutual synergy between user friendliness and transparency that should be exploited by the manufacturers and service providers to better satisfy future users. Also, network and terminal homogeneity are concepts that support each other naturally.

Figure 2.3 [1] shows the innovation areas introduced by previous, present, and future mobile communication networks as well as the associated open system interconnection (OSI) layers to these innovations. It is expected that 4G would require innovation across all OSI layers. One can see also that the decision of a user to buy a mobile terminal is chiefly related to what the upper layers

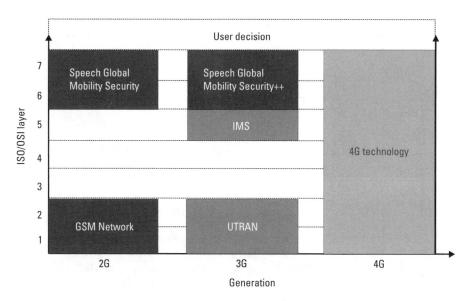

Figure 2.3 Innovations in different ISO/OSI protocol layers in different mobile communication generations.

have to offer (e.g., applications and services attainable through the terminal), rather than other obscure aspects (for the user) like physical layer features.

From the viewpoint of market trends, it is an unchanging and everlasting truth that markets always follow the economic costs and profits, no matter how many new high technology solutions are developed. As of today, high-speed data connection is a norm in accessing multimedia and Internet services through wired connections. This is also starting to become true for wireless environments, due to additional spectrum, efficient spectrum utilization, and advanced radio technologies. Services are the driver for the future market and the role of technology is just to enable provision of services and capabilities for which users are willing to pay. Figure 2.4 shows the predicted worldwide revenue for different 3G services until 2010. By that time services based on data traffic may contribute with more than two-thirds of the total service revenues. The global geographical distribution in Figure 2.5 shows that Asia and Europe will enjoy the largest revenue shares in mobile data communications. Today the existing mobile subscriber base is huge, around one-quarter of the world's population. Such an impressive figure speaks by itself in terms of potential for an explosive growth in traffic.

Even though 4G is approached from a user-centric perspective, rapidly changing technology trends need to be taken into account when developing future scenarios. Technology-wise, in the ITU-R M.1645 recommendation [3], technologies and trends are related and the key examples are provided in five

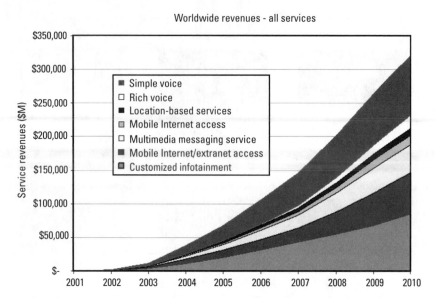

Figure 2.4 Global revenues from mobile voice and data communications (only 3G).

categories: (1) system-related technologies, (2) access network and radio interface, (3) utilization of spectrum, (4) mobile terminals, and (5) applications. Taking into consideration the above trends as a starting point, future trends shown in Table 2.1 were obtained by using a well-known expert agreement approach (a.k.a. Delphi method). We evaluated each small trend as to its significance in future society.

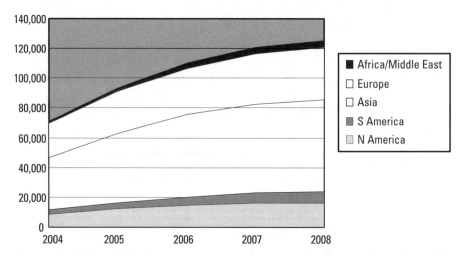

Figure 2.5 Geographical distribution of revenues from mobile data communications.

Table 2.1
Future Trends

User	Technology	Economy & Industry	Politics & Society
* Increased interest in environment * Increased interest in health * Increased user driven service * Increased demand for information * Increased demand for personalized services * Interaction within user group becomes more visible (e.g., enabling cooperative communications)	* Environment friendly technology development * User interface, intelligent agent technology development (e.g., avatar) * Increased system complexity (Reed's law) * Open architecture dominant * Increased software dependency	* Newly industrialized countries (NIC) activeness (e.g., BRICs) * Market competition activeness (efficiency stressed competition system) * Broadcasting communication integration * Increase in the need for cost efficient convenient network application	* Spectrum regulation enforcement * Environment regulation enforcement * Market dominance transfer regulation * Transfer to knowledge society * Increased demand for public safety network

2.3   A User-Centric Approach to Developing Services

In this section we will present a framework that aims to develop new services by taking the user as a starting point. As an intermediate step to *new service development* we will focus first on *scenario development*. In Section 2.4 we will concentrate on developing 4G scenarios. In Section 2.5 we will develop some service concepts for the proposed scenarios. Authors in [4] suggest three different ways for scenario development: normative, inductive, and deductive. The normative and inductive methods start from a participant's assumptions on the future society, but the results can be biased towards participants' preconceptions. On the other hand, the deductive method starts from the concrete facts, eliminating the participants' preconceived ideas as much as possible, which is why many of the existing documents have been adopting it. Readers are referred to [4] for more detail on normative and inductive methods. We summarize our framework as shown in Figure 2.6.

2.3.1   Scenario Development

Instead of vague and abstract questions like, "What will society look like in the year 2010?" we would like to ask a more concrete and specific question: "What

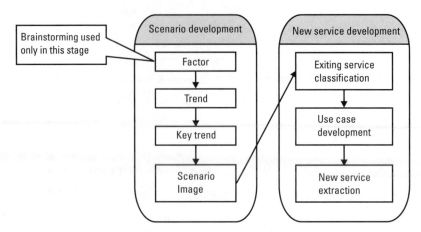

Figure 2.6 Scenario and service development methods (deductive).

will the mobile device in 2010 look like?" or "What kind of killer applications will mobile devices offer in 2010?" These allow us to stay more focused.

(1) Factors

A. *Identification:* Participants brainstorm to identify various factors that may influence an answer to the question. We will call discrete factors that remain undivided "factors." We believe that any systematic framework needs at least one phase where it needs participants' input, and it should be the first stage of the entire process. From then on, the process should be executed in a relatively automatic and objective way from the first stage on. Thus, our framework starts with a brainstorming section as in other similar work. Unlike others, however, our framework does not rely as heavily on brainstorming or common wisdom after the first brainstorming.

B. *Classification:* The list of factors will have much duplication and will include similar or overlapping ideas because of the nature of brainstorming, which encourages free and unobstructed thought. In order to facilitate the removal of duplication or overlaps, we first classify the factors.

C. *Evaluation:* We evaluate how important each factor is by using the Delphi method [5].

(2) Trends

A. *Identification:* We group the factors that have similar impact into clusters (classification of the factors helps here). We call these clusters trends. If a factor in a trend drives (prohibit resp.) the trend, we call the trend an activator (inhibitor resp.) of the trend.

B. *Evaluation:* We evaluate each trend's importance and uncertainty based on its activator's and inhibitor's importance. We believe that this new method gives more reliable results than directly evaluating a trend's importance and uncertainty by using the Delphi method, mainly because trends are too big to understand solely by intuition.

(3) Key Trends

A. Selecting the scenario logic (matrix): Depending on the number of important and uncertain trends, we can proceed in one of two ways:

i In the case where we can identify two or three important and uncertain trends [4]: We set two or three possibilities that each trend may assume. For instance, a trend that "environmentalism will become a major concern," can have two different possibilities, namely, "high" or "low." It is up to the participants' judgment to allow a third possibility: "somewhere in between." Then we can construct a matrix that spans all the possible situations (we dub it a scenario quadrant) depending on how each important and uncertain trend turns out. Then we exclude scenario seeds that seem very unlikely (see Figure 2.7 and Table 2.2).[1]

ii In the case that we can identify two or three important and uncertain trends [6]: We set two to five possibilities that each trend may assume. Unlike the case above, we have too many possible combinations. Thus, we choose a few likely scenario seeds and make a judgment on how each important and uncertain trend will turn out.

(4) Scenario Images

A. Choosing scenario plots: We select an appropriate plot for each scenario seed. We adopt a pool of stereotypical scenario plots (that is, those that occur again and again throughout the history) from [6]: winner and loser, challenge and response, evolution, and so forth. We characterize each scenario plot by a set of attributes and apply the same set of attributes to each scenario seed. Then we match each scenario seed to a scenario plot that best matches with respect to the attribute sets. We believe that this systematic method gives more reliable and consistent results than selection by common wisdom.

---

1. Selected trends should be as orthogonal as possible in order to span the whole space of possibility in the question. The fact that there is an unfeasible combination of the selected big trends implies that they are not really orthogonal. In that case, we may need to scrutinize the selection process.

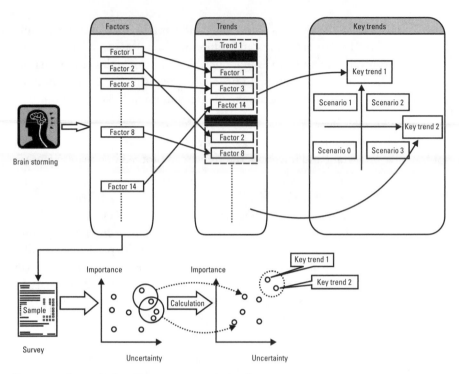

Figure 2.7 Scenario development process in detail.

B. Drawing overall images: From the scenario seeds and the selected scenario plots, we describe an image of the future, which we dub the overall image.

Table 2.2 shows the selection of the scenario logic. Scenario 1 occurs when Trend 1 takes Possibility 1, Trend 2 takes Possibility 1. Scenarios 2 and 3 occur under the same rule. The possibility that Trend 1 takes Possibility 2 and Trend 2 takes Possibility 1 is very unlikely, so it is deemed unfeasible and will not be considered.

Table 2.2
Scenario Logic

	Key Trend 1	
Key Trend 2	Possibility 1	Possibility 2
Possibility 1	Scenario quadrant 1	Unfeasible
Possibility 2	Scenario quadrant 2	Scenario quadrant 3

2.3.2 Service Development

From the scenarios previously made, we develop new services in the following manner:

1. Classifying existing services: We classify existing services gleaned from various sources in terms of a set of attributes and evaluate how promising or suitable each service will be in each overall image.

2. Use case development: Given the overall images, and the promising services, we flesh out scenarios; this method is in sharp contrast to other approaches in the available literature, where scenario creation relies on common wisdom.

3. New services extraction: We derive new promising services for each scenario. In doing so, we believe that the overall image and promising services will help.

2.4 Scenarios

Factors, trends, key trends, and scenario images are discussed in the following sections.

2.4.1 Factors

Through various documents and recent data we have extracted and classified factors related to future trends from the user, technology, economy, and industry standpoints. These results were shown in Table 2.1. By using an expert agreement method, we evaluated the impact of each factor on future society.

2.4.2 Trends

Classified factors are clustered suitably. If all the factors in a cluster facilitate specific trends, then those factors in the cluster are the activators of those particular trends. From the importance of factors acquired from the previous step, we acquire the importance of trends as follows: If there are many important activators and important inhibitors and the difference between them is small, then that trend is important and uncertain. These trends become the main axis in deciding the scenarios. On the other hand, if important activators or inhibitors are many and the difference is large, then that trend is important and certain. The other trends are not considered from then on. Table 2.3 shows the list of trends and their activators/inhibitors.

Table 2.3
The Trends

1. Users will become more active.
2. As basic needs are fulfilled, people begin to search for deeper values of life.
3. The craving for personalization will grow.
4. Technology will pervade ubiquitously.
5. Environmentalism will become a major concern.
6. Discrete technologies of the digital divide shall merge towards digital convergence.
7. Globalization shall increase.
8. A technology-abundant society consisting of cyber lives creates a demand for greater security.
9. Value chains will increase in complexity.
10. Global aging of societies shall increase.

2.4.3   Key Trends

The trends can be classified in the following three ways by using the Delphi method: Each expert ranks the trends by importance and uncertainty and their results are aggregated to reach a consensus as shown below (see Figure 2.8).

(1) The two most important and uncertain trends — will be the key trends.

   A. User activeness shall increase (user activeness)

      i. Activator

         1. Increase of user activeness on information application (user role changes from contents consumer to producer);

         2. Increase of demand of large amount of multimedia data (game, mod) (transfer to knowledge society);

         3. Increase in user driven service.

      ii. Inhibitor

         1. Increase in system complexity;

         2. Increase in demand for intelligent life environment: pleasant and convenient life.

   B. People begin search for the value of life (value of life)

      i. Activator

         1. Increase in demand for intelligent life environment: pleasant and convenient life;

    2. Leisure time and personal experience value increase;

    3. Increase in demand for user-friendly system (mutual communication based on five senses).

  ii. Inhibitor

    1. Transfer to efficiency economy important system due to market competition activeness (do not consider external cost such as environment and consider explicit internal cost);

    2. Infinite increase in information and the increase in the problem of choice.

(2) The runner-up could be the additional key trends in case we need to more finely divide the scenario images.

A. The craving for personalization will grow.

B. Ubiquity shall pervade in to all technology fields.

(3) Other trends will be given or assumed on hold.

2.4.4 Scenario Images

Plots are basically the structure of a scenario or the outline of a large framework. Plots must have clear causality and must not have contradictions overall. The characters in these plots are either driving forces or institutions. The scenarios describe how these driving forces or institutions might behave under different plots or combinations of plots. Reference [6] shows the following stereotypical plots. Note that this is one possible way to classify stereotypical plots, although this approach [6] widely used. Thus, as an example for organizing, classifying, and analyzing scenarios, we consider the following.

1. Winners and losers: Serious complications and secret sympathy is the outlook of the world. More than just what I want, it is important to know what opposes me or who opposes my enemies. Basically, this is a zero sum plotline where the strong survive and the weak get weaker.

2. Challenge and response: Conquer difficulties and make technological or political breakthroughs.

3. Evolution: Gradual changes, forecast based on past trends. When considering technical advancement, a technology map is important, especially technology advances that affect human life.

From the scenario seeds extracted from the previous stage we choose suitable scenario plots, based on which we picture overall images of the society (see Figure 2.8).

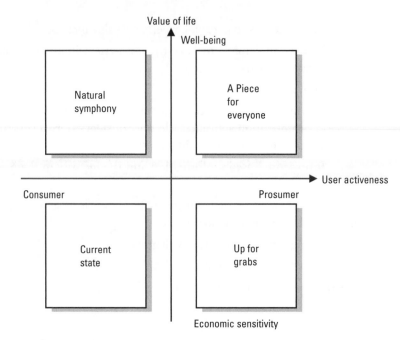

Figure 2.8 Scenario dimensions.

Natural Symphony

Key phases: ubiquitous network, value of life.

This scenario shows a society where constituents are well connected via ubiquitous network. Users pursue their lives with the full assistance of various mobile services, which provide push service to users, automatically recognizing users' needs.

Market. In this ubiquity society, there exists competition among different business areas such as mobile operators, broadcasting operators, and access network providers, but this is very reasonable competition within a few dominant market leaders that shows no threats of cartel. Big moguls with affiliated companies like mobile operators, service providers, and access network providers could offer seamless and personalized services with qualified levels to users.

Society. This society is very comfortable and peaceful with a low crime rate. Due to ubiquitous technologies, many things (e.g., security) that people had to worry about are resolved. After that, people have a greater interest in health, travel, and so on, which speaks to valuable lives. As the personalized push service becomes more universal, individualism is more and more apparent in this society.

A Piece for Everyone

Key phases: Community, diversification.

This scenario shows society as it gradually changes in time without any great turbulence. The users in this society are very active and consider personal happiness very important. DIY-like activity is quite common with full automation or ubiquity partially deployed. Because of user activity patterns, unlicensed spectrum usage is very high and research on spectral efficiency is highly motivated.

Market. Business fields such as services, access networks, and network management are run independently of each other, and the network architecture takes the form of open networks. Personalized services are not enough for the active users, which makes them prosumers, making the access network more important than services. Due to the extensive amount of networks, convergence has only been partially realized.

Society. As people become more active in the pursuit of personal happiness, society becomes an enormous set of unique communities according to hobbies and interests. Communities are very specialized and detailed in all fields, but due to their heterogeneity they lack commonality between them.

Up for Grabs

Key phases: It is economics stupid, merciless jungle, winner and loser.

This scenario implies that society evolves extremely biased to capitalistic economy values and everyone seeks to accumulate economic wealth, neglecting other values of life. Most enterprises only provide uniform products and services in order to satisfy similar user needs.

Market. User demand will concentrate on specific services and products related to economical behavior. Users may always evaluate services and products in search of cost-effective methods to use or produce valuable information. Simple information providers may lose their market power since most users actively collect information by themselves to produce economic values. On the other hand, access network service market will grow and industrial structure of access network service market may converge to oligopoly due to the large investment necessary to construct a core network. But, due to such high competition, any access network provider cannot become a dominant market power.

Society. Economic value is most important to most members of society. Most members may neglect noneconomical behavior such as human relations and the dilettante life; hence, a person or organization will be judged according to the economical value of his outcome. This extreme capitalism will cause noticeable distinction between the rich and the poor. Population will crowd towards downtown and the outskirts. Workers will usually go to work by mass transportation.

In this section we have shown that by following a reasonable and rational analysis it is possible to identify key scenarios that will serve as a starting point to develop appropriate services and applications.

2.5   Services

After considering *users* and identifying promising *scenarios* in previous sections, in this section we move on to discuss *services* for future wireless communication systems. Services cannot be developed in isolation; they need to be conceived around specific scenarios. We first start with a classification of existing services in Section 2.5.1. Then, in Section 2.5.2 we concentrate on service extraction.

2.5.1   Existing Services Classification

We examine each service presented in various sources in terms of a set of attributes and determine whether each service is suitable or promising in the three overall images (See Table 2.4).

2.5.2   New Services Extraction

We extract new services from the use cases we developed in the previous stage. Figure 2.9 corresponds with Natural Symphony. Figure 2.10 corresponds with A Piece for Everyone. Figure 2.11 corresponds with Up for Grabs.

We have shown how our framework makes full use of participants and produces cascading results that eventually lead to complete scenarios. Our framework leverages participants' input gleaned at the beginning; it then utilizes and leverages it to the full extent throughout the entire process. In our view, this framework outperforms the existing ones with respect to efficiency, consistency, and reliability.

We believe, although it is still to be verified, that our framework can be a basis for developing new services; our framework provides an evaluator that determines how promising a proposed service is in a particular scenario. Thus, we may be able to develop a new service and constantly test it using the evaluator, which would make the new service more solid and profitable. WWRF [16] and mITF [15] (see Chapter 5) have also developed scenarios and services in an attempt to justify the development of future wireless communication systems. Surprisingly, one can find several similar concepts and proposals for future services.

2.6   Visions and Foresights

In this section we move on from the previously discussed user-scenario-service chain, and approach 4G from a more technical perspective. We first review the ITU vision of future wireless communication systems, considered here as our baseline reference. Next, we describe and discuss some common approaches to 4G, trying to identify some commonalities and differences. Finally, in an attempt

Table 2.4
Service Examination with Respect to Attributes, and Suitability for Each Image

Service	Attributes					
	Place	Prosumer	Interactivity	A Piece for Everyone	Up for Grabs	Natural Symphony
Mobile administrative service	Anywhere	No	Yes			✓
Voice calling	Anywhere	Yes	Yes	✓	✓	✓
Video calling	Anywhere	Yes	Yes	✓	✓	✓
Info mobile	Building	No	Yes			✓
Broadcasting	anywhere	No	No	✓	✓	
Group study/education	School/home	Yes	Yes	✓	✓	
Remote education	Outdoor	Yes	Yes	✓		
On-demand knowledge center	Anywhere	No	No	✓		
Travel agent	Anywhere	No	No			✓
Data and multimedia delivery service	Anywhere	No	No			✓
Mobile game	Anywhere	No	No	✓		✓
MM conference	Anywhere	Yes	Yes	✓	✓	✓
Food manager	Home	No	Yes	✓		✓
Interpreter	Anywhere	No	No	✓		✓
Personal assistant home agent	Home	No	Yes	✓		✓
Personal mobile assistant	Anywhere	No	Yes			✓
M-ticketing	Anywhere	No	Yes			✓
Security assistant home agent	Home	No	No	✓	✓	✓
News/weather notification	Anywhere	No	No	✓	✓	✓
Spot info push	Anywhere	No	No			✓

Note: Not all the services and attributes are shown here in order to save space.

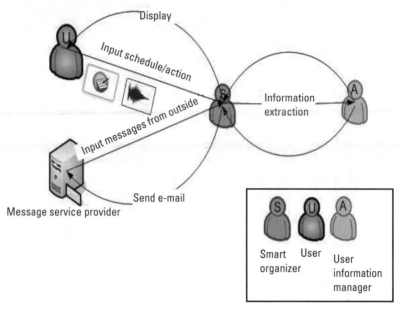

Smart organizer

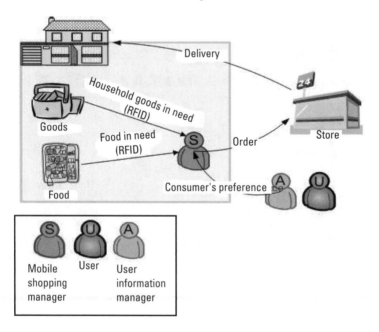

Mobile shopping

Figure 2.9 (a) New services extraction for Natural Symphony.

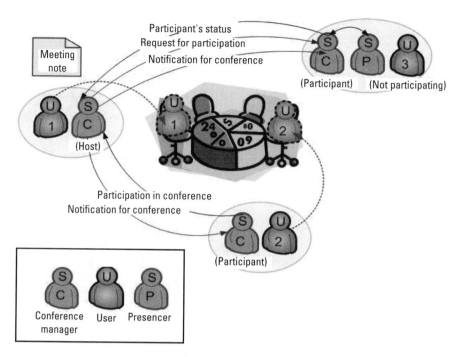

Conference manager

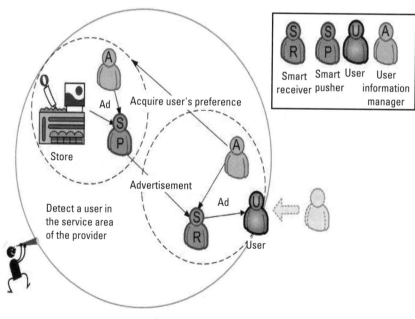

Smart pusher/receiver

Figure 2.9 (b) continued.

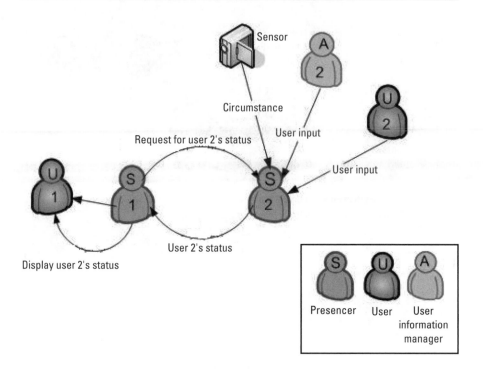

Presencer

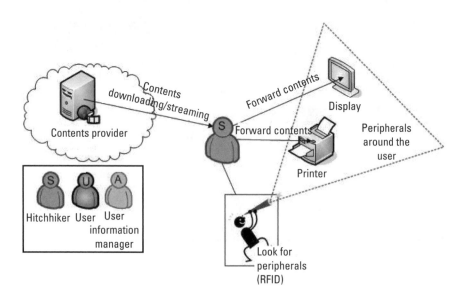

Hitchhiker

Figure 2.9 (c) continued.

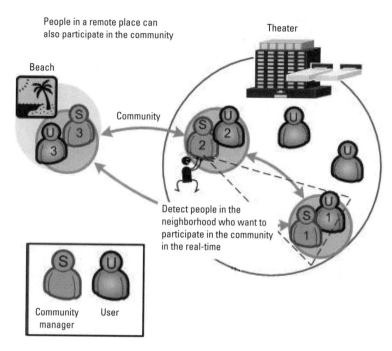

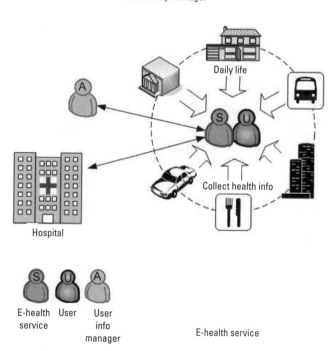

Figure 2.10 New services extraction for A Piece for Everyone.

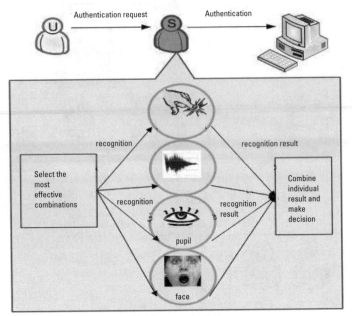

Unified biometric authentication

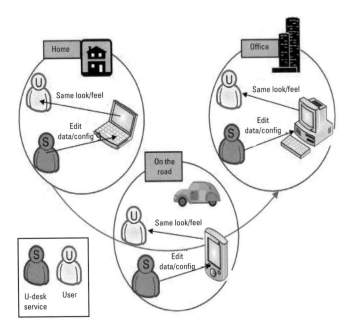

U-desk

Figure 2.11 (a) New services extraction for Up for Grabs.

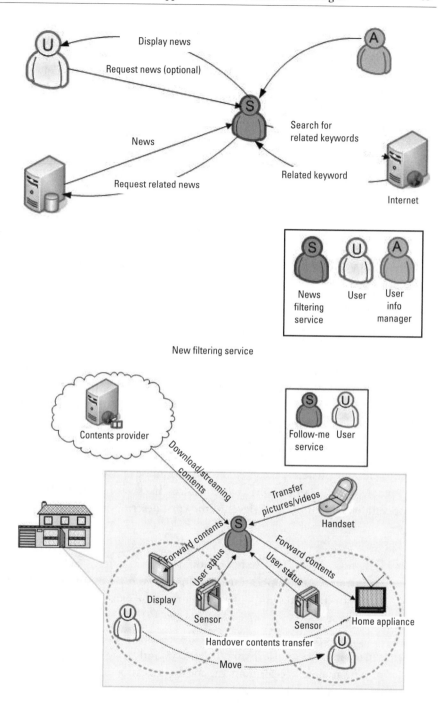

New filtering service

Follow-me service

Figure 2.11 (b) continued.

to foresee 4G, we present our own views. In addition to the user trends previously discussed, other important developments and tendencies could help us to better approach, define, and justify future 4G networks. Trends in various categories of Table 2.1 support the argument of an explosive increase in the demand of bandwidth required to realize a truly *wireless knowledge society*. Current wireless systems like 3G fit very well in this picture, but only as one constituent part. Present systems cannot by themselves offer enough network capacity, link capabilities, and coverage for supporting a vision of a highly connected world fully exploiting broadband communications. Moreover, the bandwidth required for that purpose is far beyond the figures in use today, and hence, a new spectrum allocation is necessary. The vision of a highly connected world, where virtually every entity (users and machines) can communicate to each other in any possible situation, anytime, exploiting broadband capabilities, is truly appealing, and equally important, fits very well in the aforementioned logical chain.

2.6.1   ITU Vision

The ITU-R Working Party 8F (WP8F) is responsible for the overall system aspects of IMT-2000 and beyond. The ITU developed the 4G vision not only for internal ITU work but also for external organizations in order to provide an official and unbiased overview of 4G for related industries, the global research community, and other interested parties. Note that ITU uses the terminology "beyond IMT-2000" rather than 4G.

The WP8F believes that the IMT-2000 (in its present state), along with further enhancements to the IMT-2000 like new mobile access and nomadic/local area wireless access elements, should be considered to be a system as a whole. It is important to consider backward compatibility and interoperability. The ITU high-level framework describes this point clearly, as follows:

- Enhanced IMT-2000 systems should support a steady and continuous evolution of new applications, products, and services through improvements in data rates and enhancements to the existing IMT-2000 radio interfaces.
- A framework for systems beyond IMT-2000 should be realized by the functional fusion of existing, enhanced, and newly developed elements of new mobile access, nomadic local wireless, and so forth, with high commonality and seamless interworking.

This concise definition of 4G is in agreement with the definition given in Chapter 1. We stress again the fact that ITU in principle identifies 4G as a system having a functional fusion of existing (legacy) systems, enhanced systems, and new capabilities with a new spectrum-though it does not establish if a new spectrum is required or not. The ITU focuses more on developing a new system rather than considering the fusion of legacy systems. Concrete standardization may commence once spectrum has been identified, presently assumed to be at the world radiocommunication conference (WRC-07), with a period of research and investigations required prior to this meeting. In order to successfully provide broadband wireless connectivity to the expected large number of 4G users and new bandwidth-hungry services, it is likely that additional and/or new spectrum must be identified. A more detailed process for spectrum identification in WRC-07 is described in Section 2.7.

Figure 2.12, developed in WP8F and sometimes referred to as the van diagram, shows the capabilities of 4G, illustrating the various components and their relationships to each other.

The capability of future 4G systems can be expressed into three parts;

1. Future development of the IMT-2000. There will be a steady and continuous evolution of the IMT-2000 to support new applications, products, and services. For example, the capabilities of some of the IMT-2000 terrestrial radio interfaces are already being extended up to 10 Mbps and it is anticipated that these will be extended even further up to approximately 30 Mbps by 2005, under optimum signal and traffic conditions.

2. New capabilities of systems beyond IMT-2000. For systems beyond IMT-2000, there may be a requirement for a new wireless access technology for the terrestrial component-expected around 2010 and after. This will complement the enhanced IMT-2000 systems and the other radio systems. ITU as well as key telecom manufacturers predict that by 2010 potential new radio interface(s) should support data rates of up to approximately 100 Mbps for high mobility such as mobile access and up to approximately 1 Gbps for low mobility such as nomadic/local wireless access. These data rate figures and their associated mobility should be seen as the targets for research and investigation of the basic technologies necessary to implement the framework. Note that 100 Mbps and 1 Gbps refer to overall cell throughput in wide and local access, respectively. Future system specifications and designs will be based on the results of the research and investigations. Due to the predicted data rate requirements, additional spectrum will be needed in order to deliver the new capabilities of systems beyond IMT-2000. The data rate figures anticipate the advances in technology, and these values are expected to be technologically feasible in the time frame noted above. It is possible that upstream (uplink) and

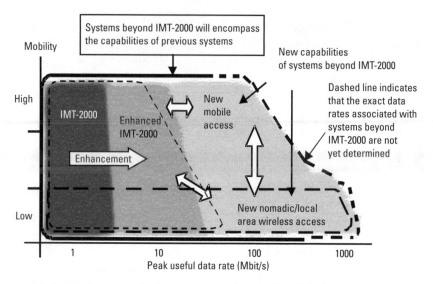

Figure 2.12 Illustration of capabilities of the IMT-2000 and systems beyond IMT-2000. (From: [3].)

downstream (downlink) may have different maximum transmission speeds.

3. Relationship between the IMT-2000 systems, beyond IMT-2000, and other access systems. In conjunction with the future development of IMT-2000 and systems beyond IMT-2000, relationships will continue to develop between different radio access and communication systems; for example, WPANs, WLANs, wireless metropolitan area networks (WMANs), digital broadcast, and fixed wireless access (FWA). As pointed out before, other communication relationships besides person-to-person will emerge, such as machine-to-machine, machine-to-person, and person-to-machine.

The ITU-T also developed the Recommendation ITU-T Q.1702, entitled "Long-Term Vision of Network Aspects of Beyond IMT-2000," which addresses the envisaged network environment, network design objectives, and architecture concepts of systems beyond IMT-2000. This ITU vision would be the nucleus and guidelines for the development of related research in external organizations as well as within the ITU.

The world today is comprised of mutually exclusive networks, which are now beginning to interwork with each other. In that sense the future can be seen as a convergence of networks. Thus, as it was already mentioned, one can interpret 4G systems as a converging platform where heterogeneous networks interoperate in a seamless manner. Moreover, from a broader perspective, the overall composite network can be considered as a network of networks. Such a convergence cannot take place without the concurrent efforts of all involved parties. Broad consensus is required across industry. Operators and service providers must be part of the business equation. Universities and other research centers play a key role not by supporting basic investigation but also by actively contributing and generating new concepts and technologies to the field. It has been recognized that 4G development should be carried out around the user, and with the user as the ultimate target. Understanding the user means matching future products and services to the market needs. No one can take 4G for granted; in order to make 4G true and thoroughly pervade our lives, an enormous amount of effort is required to find a global consensus.

2.6.2 Some Approaches to 4G

Several approaches to 4G exist. The simplest ones tend to see 4G as a linear extension of current 3G systems, particularly highlighting, either separately or jointly, the high data rate transfer capabilities and the support of high mobility. This *linear vision*, to some extent prevalent in Asia, represents a rather narrow vision, serving perhaps the interests of certain sectors, but fails to see *4G as a universal, flexible, ubiquitous, and powerful access solution working in and across a variety of scenarios and supporting a variety of services from a wide range of possible terminals*. This linear vision is also rather common in America, where most of the attention seems to focus on the high-speed capabilities of the access schemes, but in this case the focal point is on local access (short-range) systems.

A *concurrent vision* of 4G, often identified as the European vision, supports the *integrative role* of 4G as a convergence platform of several networks. In other words, one could think of 4G as a network of *heterogeneous networks*. The integration of such an eclectic system is achieved at an IP networking layer, where the role of IP is paramount to enable a seamless network operation. An all-IP network (encompassing the access and core networks) is the most straightforward and effective way to integrate all the possible different networks constituting 4G. Regardless of particular views to approach 4G and despite differences in its definition, developing parties tend to agree on several of the key characteristics of such systems. The features are listed in Table 2.5 [20].

The concurrent (integrative) approach to 4G is very close to the ITU vision considered in Section 2.6.1. From such a perspective, 4G can be understood as a mosaic of complementary scenarios and associated access technologies. For each scenario there could be more than one proponent access solution. It is interesting to note that regional schedules to adopt 4G appear to differ considerably, with Asia aiming to introduce 4G by 2010, and Europe taking a slower pace, targeting 2015 or even later. It is expected that the introduction of 4G will be carried out initially by realizing the linear vision; namely, systems supporting high-speed communications in any of the possible scenarios (WWAN, WMAN, and WLAN) would be promoted as 4G. Integrating several access technologies, complementary and competing, to realize a harmonious network would be one of the greatest initial challenges of 4G.

2.6.3 Foreseeing 4G

Present mobile communication systems have evolved by adding more and more system capabilities and enhancements. In the future, and continuing with this trend, users will see a significant increase in capability through the future enhancements of existing 3G systems. As ITU states, future 4G wireless communications systems could be realized by functional fusion of *existing, enhanced,* and *newly developed* elements of current 3G systems, nomadic wireless access

Table 2.5
Characteristics of 4G Systems

Achievable data rates: 100 Mbps (wide area coverage), 1 Gbps (local area). These are design targets and represent cell overall throughput.
Networking: All-IP network (access and core networks).
Access provision: Ubiquitous, mobile, seamless communications.
Shorter latency: Connection delay 500 ms; Transmission delay 50 ms.
Cost per bit: 1/10 to 1/100 lower than that of 3G.
Infrastructure cost: 1/10 lower than that of 3G.
Plug & Access network architecture.
Enabling person-to-person, person-to-machine, and machine-to-machine communications.
Technical capabilities defined after new services and applications are identified. Services and applications fully exploiting the technical capabilities of the system.

systems and other wireless systems with high commonality, and seamless interworking. This is a generous and flexible definition, truly allowing legacy systems (2G and 3G), their evolutionary development, and new systems to *coexist*, each being a component part of a highly heterogeneous network, the 4G network.

It can be said that future 4G systems are likely to be based on evolutionary and revolutionary technologies, the latter aspect referring to newly developed access concepts. Figure 2.13 depicts the 4G realm as a conjunction of two well-known evolutionary developments, corresponding to the mobile (wide area coverage, cellular) and nomadic (local area coverage, short-range) evolutions in the upper and lower portions of the figure, respectively. From its inception in the 1980s until its current days, wide area cellular mobile communications, characterized by high mobility and low to moderate data rates, have enjoyed a steady growth in terms of achievable throughputs. Every generation of cellular systems (2G, 3G) offered marked improvements with respect to its preceding one. The same applies with the nomadic (local area) access, characterized by low mobility and moderate to high data rates. Between 3G and 4G there is a transitional period, sometimes known as B3G, where further enhancements are expected, particularly in the achievable communication speed together with the degree of mobility. As time goes by, one can see that local access systems will tend to acquire more characteristics of wide area systems (e.g., higher mobility) whereas wide area systems will tend to become closer to their nomadic

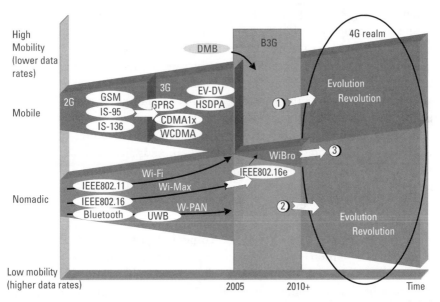

Figure 2.13 Evolution of mobile and nomadic wireless communication systems and their convergence into the 4G domain.

counterparts in terms of data throughputs. It is obvious that the capabilities of 4G have to be higher than those corresponding to any of the present wireless systems, encompassing virtually all possible combinations of mobility and data rate. The shift toward higher data rates may pose challenges to the designers of future systems, in particular when high data throughput connections and mobility are to be supported simultaneously. In addition to the (1) *mobile* and (2) *nomadic development paths* (corresponding to WWAN and WLAN developments, respectively), Figure 2.13 also shows a third path plainly aiming to (3) 4G, namely through broadband wireless access technology (sometimes identified as WMAN). This third approach has received lately considerable attention and support, and hence, it is also worth taking it into account as a legitimate technology component aiming at 4G. WMAN refers mainly to the developments around the Institute of Electrical and Electronic Engineers (IEEE) 802.16 standards. In particular, standard version 802.16e addresses the problem of mobility in mid-size cells (e.g., up to several kilometers) while maintaining relatively high data throughputs (e.g., tens of megabytes per second). In terms of mobility and data transfer capabilities, WMAN can be positioned on a middle way between WWAN and WLAN. Also, from the standpoint of these technical features, we can consider WLAN, WMAN, and WWAN as *complementary techniques* that can coexist and altogether constitute the 4G network. It is likely, however, that evolution and further developments to enhance the most important features of these networks, conferring wider coverage, more mobility, and higher data throughputs, will result in overlapping capabilities. Obviously, this would lead to scenarios where these three basic component technologies could also be competing. In such cases, it is difficult to forecast how these component technologies could be positioned, as the interaction between manufacturers, operators, and markets is complex. Infrastructure and terminal costs will have a decisive role in determining the final share of these possibly *competing* approaches. Certainly, other system performance indicators could also make a difference when comparing competing approaches.

As one would expect, telecommunication manufacturers (sometimes referred to as the mobile sector) tend to see 4G more as an evolutionary continuation of their current mobile business, and thus putting more emphasis on development falling within the upper right corner of Figure 2.13. Conversely, IT companies (the wireless sector), with background business in local access systems, are more inclined to see 4G as an enhanced extension of such short-range communication systems (the lower right corner in Figure 2.13). As already mentioned, mere evolutionary development may not necessarily allow the achievement of some goals, and some new developments could be needed. Academia, regulatory bodies, and other parties with fewer economic ties to the wireless business tend to favor a more unified vision of 4G, supporting equally both wide area and local area approaches.

At this point it is interesting to point out that though highly widespread, the terminology "4G" is somewhat misleading. Indeed, previous and present generations (1G to 3G) refer chiefly to cellular access, whereas 4G, in its widest sense, refers to a broader domain where cellular is just one of its constituent parts. Due to this and some other strategic reasons, some parties prefer to avoid the use of "4G." In particular, ITU refers to Beyond IMT-2000 when referring to the next generation of wireless communication systems, as already mentioned.

One can approach 4G from the network coverage standpoint, by looking at how different wireless services were and are being provided at different geographical scales [20], as depicted in Figure 2.14 [2]. One of the tenets mostly associated with 4G is *"always connected, everywhere, anytime."* The outcome of such a principle, though highly beneficial to the end user, demands unprecedented efforts on the part of the designers of future wireless communication systems. The apparent simplicity and transparency enjoyed by users of future 4G systems has an enormous price to be paid by manufacturers, standardization bodies, and the research community, in terms of having a network of eclectic networks to appear as a single, simple, and everywhere-reaching network. Figure 2.14 [2] shows the network hierarchy, starting with a *distribution layer* at the largest scale. This layer provides large geographical coverage with full mobility, although links may convey a chunk of composite information rather than signals from individual subscribers (e.g., broadcast services such as digital audio broadcasting (DAB) and digital video broadcasting (DVB)). Next in the hierarchy is the *cellular layer*, with typical macro-cells of up to a few tens of kilometers. This network also provides full coverage and full mobility, but now connections are intended to cater to individual users directly. Global roaming is an essential component of 2G

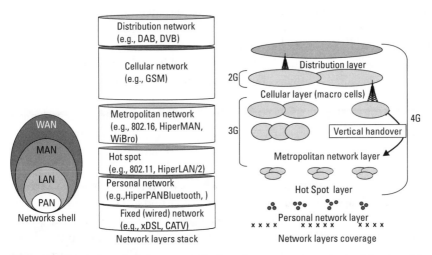

Figure 2.14 Evolution of wireless networks from the coverage standpoint: (a) network layers stack; (b) network layers coverage. (From: [2].)

cellular systems (e.g., global system for mobile communications (GSM). Note that the cellular layer encompasses both macro and micro-cells. The *metropolitan layer or network*-of which IEEE 802.16, HyperMAN, and Wireless broadband (WiBro) are typical examples-provides urban coverage with a range of a few kilometers at the most, with moderate mobility and moderate data speed capabilities. On a smaller scale and moving towards the *local area layer* (e.g., indoor networks or short-range communications), the network provides access in a pico-cell, typically not larger than a few hundred meters, to fulfill the high capacity needs of hotspots. Nomadic (local) mobility is supported, as is global roaming. 3G makes use of the cellular layer (typically micro-cells) in combination with hotspots (WLAN), through vertical handovers, to provide coverage in dedicated areas. Next in the wireless network hierarchy is the *personal area network* (PAN), very-short-range communication links (typically 10m or less) in the immediate vicinity of the user. Within this layer we can also enclose *body area networks* (BAN) and some other submeter wireless networks. Going back to the paradigm of a pervasive 4G wireless network, in order to effectively have an unlimited reach while being able to support a variety of data rates, 4G would have to embrace all the described network layers; or in other words, 4G could be defined as a convergence platform taking in and working across BAN, PAN, LAN, MAN, cellular, and distribution networks.

2.7 4G Spectrum Issues

Spectrum is a crucial matter in the design and manufacture of wireless systems and devices. The characteristics of propagation and technical parameters of the frequency bands may influence or determine requirements of devices for systems operating in the frequency bands and service characteristics such as mobility requirements and coverage. Furthermore, spectrum is a scare resource, and therefore, it should be utilized efficiently. Suitable and affordable frequency bands should be allocated to each wireless system for the purpose of efficient use of spectrum.

On the other hand, to pursue economics of scale in wireless communication systems as well as to promote the fair use of spectrum among countries, global harmonization is a very important factor to determine frequency bands for wireless services; hence, regulatory matters should be considered in the determination of spectrum allocation. This is why any spectrum of wireless services needs to be managed by a global standardization body. The ITU-R plays an important role in allocation/identification of wireless services and the WRC supported by the ITU-R is the only entity that can make decisions on the global use of spectrum. All the agreed-upon decisions in the WRC are reflected into the ITU-R Radio Regulation through the Radio Assembly (RA).

Spectrum allocation and identification by the WRC for wireless services require a considerable amount of research regarding their necessity, spectrum demands, and compatibility (or sharing) with other existing services in their preferred frequency bands. The ITU-R Study Groups are responsible for such research activity. Among the study groups, the ITU-R SG8 (8th Study Group) is in charge conducting spectrum-related research for spectrum of mobile services, whereas the ITU-R WP8F (Working Party 8F) is responsible for the 4G relevant studies.

2.7.1 Background

Research on the 4G spectrum is based on the ITU-R vision framework of future development and systems beyond IMT-2000, which advocates the vision of future mobile systems based on multiple radio access networks incorporated in one core network. Service aspects in terms of mobility and transmission rate have been described in ITU-R Recommendation M.1645. As per the vision of systems beyond IMT-2000 in M.1645, new mobile systems beyond IMT-2000 (4G mobile systems) require up to 100 Mbps per link with up to 250 km/h of user speed.

To meet the technical and service objectives of 4G mobile systems, more spectrum is required than that already identified for IMT-2000 and pre-IMT-2000. Therefore, the ITU-R WP8F seeks new spectrum in terms of bandwidth (spectrum amounts), affordable frequency bands, and associated sharing or compatibility problem, based on the decision of the WRC-2003 Resolution 228 (Rev.WRC-03). The WRC-07 agenda item 1.4 is to consider frequency-related matters for the future development of the IMT-2000 and systems beyond IMT-2000, taking into account the results of the ITU-R studies in accordance with Resolution 228. It states the following:

1. Considering the current IMT-2000 standards, the future development of IMT-2000 and systems beyond IMT-2000 described in the ITU-R Recommendation M.1645 and the needs of developing countries in development and implementation of mobile radio communication technology.

2. The ITU-R seeks frequency bands affordable to provide services described in M.1645 as well as spectrum requirements (required bandwidth).

3. It is responsible for sharing/compatibility between existing and future mobile services (4G) in the potential frequency bands.

In addition, WRC-03 considered that ITU-R should develop a harmonized time frame for common technical, operational, and spectrum-related parameters of systems, taking into account relevant IMT-2000 and other experience. With the considerations and notes of WRC-03's Resolution 228, ITU-R WP8F investigates spectrum matters related to 4G.

2.7.2   Plan and Timeline of Global 4G Spectrum Allocation

The Plan for the 4G spectrum allocation should follow the ITU-R WP8F schedule. The ITU-R WP8F is composed of three major Working Groups (WGs) and each WG is composed of several Sub-WGs.

Spectrum research within the ITU-R WP8F is performed in three separate directions. The first WG studies future markets for 4G. Traffic demands from future market perspectives will greatly influence the amount of spectrum required by 4G. The second WG concentrates on frequency bands and sharing/compatibility issues regarding spectrum requirements from future market perspectives. Technical and regulatory matters will also be taken into account in association with the possible frequency bands. The third WG focuses on future 4G systems figures. These figures will be represented as spectrum efficiency in each radio propagation environment corresponding to cell deployment scenarios such as macro and micro-cell deployment.

Working methods of the ITU-R WP8F pursuing the 4G spectrum are as follows:

- Identify potential future services for 4G and calculate associated traffic demands.
- Develop methodology for calculating spectrum requirements for 4G and estimate required spectrum amounts (bandwidth) for 4G that reflect future traffic demands.
- Identify candidate frequency bands for 4G mobile and nomadic systems, respectively, and conduct sharing and compatibility research in the identified bands.
- Develop technical and regulatory framework/requirements for 4G with regard to sharing/compatibility studies.
- Derive frequency bands for 4G and the required bandwidth in each frequency band.

Figure 2.15 shows the output of the ITU-R WP8F studies and their final due dates.

All the research on the WRC-07 agenda items within the ITU-R should be finalized before the conference, and normally, the results of the research conducted by the ITU-R SGs will be submitted to the Conference Preparatory Meeting (CPM) prior to the conference. Research results include the proposed

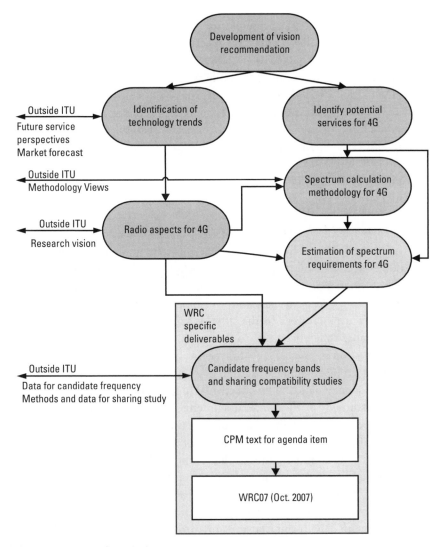

Figure 2.15 The ITU-R work plan.

frequency bands as well as bandwidth requirements in each frequency band. Based on these submitted results, the CPM will prepare CPM texts as input documents of the WRC-07.

2.7.3 Prospective Frequency Bands

Recently, discussions on frequency bands of new mobile aspects of 4G have been held in some global standardization bodies. In particular, the ITU-R WP8F deals with possible frequency bands of the 4G as well as a basic view of the 4G spectrum. The basic principles of 4G spectrum allocation within the ITU-R WP8F are as follows. The 4G spectrum should be sufficient to satisfy future market needs around 2010 and after. It should also provide full mobility and full coverage to replace current mobile systems such as pre IMT-2000 (2G) and the IMT-2000 (3G). A technical overview of the 4G spectrum will outline candidate frequency bands that can meet such requirements. In addition, existing systems should be taken into account in terms of sharing and compatibility with future 4G systems. The ITU-R allocation can provide basic information on which types of services/systems can be concerned with the future spectrum of 4G. We now review the ITU-R Radio Regulations regarding global allocation of spectrum.

Technical Aspects of 4G Spectrum

Many investigations on frequency bands for new mobile systems beyond IMT-2000 concentrate on frequency bands below 6 GHz. Several technical and regulatory aspects support frequency bands below 6 GHz for mobile systems.

First, mobile systems usually suffer from large Doppler shift as the operating frequencies are increased. In the time domain, large Doppler shift in spectrum gives smaller coherence time, so that, more frequent fading may occur in a certain transmission interval [7]. A shorter time frame, which causes an increase of overhead in transmission, is desirable to guarantee coherent detection as increase of frequency bands. The effects of Doppler shift on the future mobile systems were studied in [8]. These documents assumed that the frame size is determined in association with coherence time (i.e., frame size is defined as 20% of coherent time and size of overhead in each frame is fixed regardless of frame size). In this assumption, they showed that Doppler shift influences systems performance in terms of transmission efficiency and payload ratio during certain time intervals. If the operating frequency is large, the Doppler shift is increased. Thus, frame size should be shortened, thereby increasing the overhead.

Second, path-loss in the transmission also influences system performance and coverage, and hence, system deployment cost. Path-loss is closely related with coverage of mobile systems, and systems characterized by wider coverage work normally in lower path-loss environments. Path-loss has an explicit relation with the frequency bands. As per free-space path-loss law, path-loss without shadowing and multipath fading is inversely proportional to the frequency band. Several studies assessing the path-loss law in practical environments have been conducted. Normally, there are many obstacles that cause more scattering, diffraction, and therefore attenuation in practical environments. But, research at NTT DoCoMo [9, 10] has shown that even in urban environments, path loss without shadowing and multipath fading follows free-space law when increasing the frequency bands. Even though systems operating in low path-loss environments require longer reuse distances while meeting constant carrier-to-interference radio (CIR) requirements, more bandwidth is necessary to meet the same traffic demands in this environment than with higher frequency band environments. From this perspective, technologies that exploit shorter reuse distances are also expected to be important in the future. Lower frequency bands are more appropriate with respect to coverage, reducing eventually system deployment cost; whereas higher frequency bands are more appropriate from the standpoint of spectral efficiency.

As we described above, the appropriate frequencies for future mobile communications should be in relatively low frequency bands. To obtain bandwidth sufficient enough to support services with very high data rates, frequency bands for future mobile networks need to be above frequency bands allocated for IMT-2000-that is, above 2.7 GHz. At this point in time, there is no conclusive technical evidence on what the upper frequency limit for the band should be. However, 6 GHz appears to enjoy support among the active members of ITU working on spectrum identification and planning. Thus, as of today, frequency in the band from 2.7 to 6 GHz could be regarded as one of the most promising frequency bands for future mobile communication systems.

Regulatory Aspects of 4G Spectrum

To identify regulatory aspects of the 4G spectrum, we first review the global allocation of the spectrum below 6 GHz. The mobile spectrum below 6 GHz is mainly allocated not only to mobile services including pre-IMT-2000 (2G) and the IMT-2000 (3G) services but also to fixed services such as point-to-point or point-to-multipoint transmission. Broadcasting and broadcasting auxiliary services also use 1-GHz frequency bands extensively. Outside broadcasting normally uses 2- and 3-GHz frequency bands. The lower part of the ultrahigh frequency (UHF) band is allocated for other kinds of mobile services such as the Private Mobile Radio (PMR)/Trunked Radio Services (TRS). The terrestrial trunked radio (TETRA) and integrated digital enhanced network (iDEN) systems are being operated in these bands. Table 2.6 shows mobile allocations in the ITU-R radio regulation of the UHF band.

Some concerns on lower parts of the UHF band, below 800 MHz, have been discussed within the ITU-R WP8F regarding positions of developing countries. As per the description of technical aspects, lower frequency can make systems provide larger coverage. In developing countries, this characteristic reduces the deployment cost of mobile system with full coverage; hence, they prefer these bands for new allocation of the IMT-2000 and systems beyond IMT-2000.

Table 2.6 implies that many parts of the UHF band have already been allocated to mobile services and that most of the allocated bands are currently used or planned to be used by mobile systems including 2G and 3G mobile services. Additional identification of 4G systems in the bands allocated to mobile service

Table 2.6
Global Allocations for Mobile Services in the UHF Band

Frequency Bands	ITU-R Allocations	Operating Mobile Systems (Current and Future)
406-430 MHz	Mobile, fixed Europe: PMR U.S.: PMRS	TRS including TETRA, iDEN
440-470 MHz	Mobile, fixed Radio location (440-450 MHz) Radio sonade (460-470 MHz)	Cellular (NMT450), TRS including TETRA, iDEN
740-806 MHz	Mobile, broadcasting	No current mobile services
806-960 MHz	Mobile (IMT-2000 additional bands)	Cellular including CDMA (IS-95, 1x), GSM TRS including TETRA, iDEN
1,710-2,025 MHz 2,110-2,200 MHz	Mobile (IMT-2000 additional bands) IMT-2000 core bands (1,885-2,025 MHz and 2,110-2,200 MHz)	PCS1900, GSM 1800, IMT-2000 including FDD, TDD, and satellite component, DECT (Europe, 1,880-1,900 MHz)
2,300-2,400 MHz	Mobile, fixed	WiBro (Korea), TDS-CDMA (China)
2,400-2,500 MHz	Mobile (ISM bands)	IEEE 802.11b/g, Bluetooth
2,500-2,690 MHz	Mobile, fixed, satellite (BSS: 2,605-2,655 MHz, MSS: 2,500-2,520 and 2,670-2,690 MHz)	IMT-2000[2]

among the UHF band seems to be a rather unachievable task. Besides, other parts of the UHF band are used by other important services such azs aeronautical mobile/operations, radio location, space research, and conventional broadcast radio. Allocation of mobile services into such frequency bands is impossible because the existing services in that band would be prone to be affected by

---

2. The bands 2,500 to 2,690 MHz were identified as IMT-2000 additional bands by the WRC-2000, and use of the bands has been discussed within the ITU-R WP8F to developglobally harmonized frequency arrangements of the IMT-2000 systems. Most of the countries in the European Committee have determined that these bands be used by the IMT-2000 from 2008. The United States FCC has also announced the decision to use these bands in June 2004, and they will allow commercial mobile services including the IMT-2000 in the bands.

interference from mobile services. Therefore, it can be concluded that the UHF band is not appropriate for mobile services with regard to regulatory aspect.

Regarding those two aspects of the 4G spectrum, affordable frequency bands can hereby be concluded as the 3- to 6-GHz band. Current allocation of the 3- to 6-GHz band is shown in Table 2.7.

Candidate Frequency Band and Global Research Activities

Spectrum allocation/identification of new systems will be globally determined based on technical research and regulatory aspects. New identification can have a significant influence on the wireless communication industry and other key factions; thus, it should meet the needs of as many private and public organizations as possible. Common understanding and views on the 4G services are the primary assumption for achieving globally harmonized frequency bands for the future 4G communication systems. Technical research can provide not only

Table 2.7
Global Allocation in the 3- to 6-GHz Band

Frequency Bands	ITU-R Allocations	Current Operating Mobile Systems
3,100–3,300 MHz	Radio location Earth exploration satellite (Active) Space research (Active)	Airborne Warning and Control Systems (AWACS) Synthetic aperture radars (SARs)
3,300–3,400 MHz	Radio location	AWACS SARs
3,400–3,600 MHz	Fixed, mobile, fixed satellite, radio location	Broadband wireless access (BWA) such as WiMAX, 802.16 European countries: electronic news gathering (ENG)/Outside broadcasting (OB)
3,600–4,200 MHz	Fixed, mobile, fixed satellite	Fixed Satellite Services (FSS), FWA, including microwave link
4,200–4,400 MHz	Aeronautical radionavigation	
4,400–4,500 MHz	Fixed, mobile	Point-to-point microwave link
4,500–4,800 MHz	Fixed, fixed satellite, mobile	Global geostationary Earth orbit (GEO) satellite bands,[3] point-to-point microwave link

---

3. See WRC-07 Agenda Item 1.10.

Table 2.7
continued

Frequency Bands	ITU-R Allocations	Current Operating Mobile Systems
4,800-4,990 MHz	Fixed, mobile, radio astronomy	Point-to-point microwave link
5,000-5,150 MHz	Aeronautical radionavigation	
5,150-5,350 MHz	Aeronautical radionavigation, fixed satellite (Earth-to-space), mobile[4]	5-GHz wireless LAN[5]
5,350-5,470 MHz	Earth exploration, radiolocation, space research, mobile	5-GHz wireless LAN
5,470-5,650 MHz	Maritime radionavigation	
5,650-5,725 MHz	Radio location, space research, mobile	5-GHz wireless LAN
5,725-5,825 MHz	Radio location, amature	5-GHz wireless LAN, dedicated short-range communication (DSRC)
5,825-6,000 MHz	Fixed, fixed satellite (Earth-to-space), mobile	Broadcasting relay

solutions for the frequency bands that are affordable with respect to the characteristics of radio transmission, but also it can help to show how future systems can be compatible with the existing services in the same band and adjacent bands.

Spectrum for 4G should be identified based on following principles:

- Regarding technical aspects, frequency bands for 4G will be lower than 6 GHz.

- Regarding regulatory aspects, frequency bands for 4G will be higher than current identified bands for mobile services (i.e., 2.7-GHz bands).

---

4. Mobile use of these bands is identified by WRC-03 with regard to devices that emit only low level emission.

5. Use of WLAN in the bands 5,150 to 5250 MHz is restricted to indoor transmission with maximum effective isotropic radiated power (EIRP) of 200 mW/20 MHz, but use of WLAN in the bands 5,250 to 5,350 and 5,470 to 5,725 MHz are allowed for outdoor transmission with maximum EIRP of 1W/20 MHz.

- Regarding regulatory aspects, frequency bands for 4G will be among the bands that have already been allocated to mobile services. New allocation of mobile services is not allowed according to the WRC-03 Resolution 228.

- Regarding the needs of developing countries, lower frequency bands than those currently identified for mobile services will be used (i.e., below 800 MHz can be examined).

The bands 3,400 to 4,200 MHz and 4,400 to 4,990 MHz are possible frequency bands for 4G in consideration of the above statements. Also, the 2,300- to 2400-MHz band, already identified by mobile services in some countries such as WiBro (Korea) and time division synchronous CDMA (TDS-CDMA) (China), can be a candidate for several countries. The 5-GHz band is appropriate from the technical standpoint, but it has many restrictions of emission due to protection of existing services such as radio location (Radar) and Earth exploration satellite; hence, a cellular type of service in that band does not seem to be possible.

Such candidate frequency bands will be discussed within the ITU-R WP8F. Recently, the WP8F distributed a detailed survey document to each administration to examine current mobile services and possible frequency bands identified as spectrum for 4G. Decisions on candidate frequency bands will be associated with spectrum requirements for the 4G (i.e., bandwidth).

Spectrum assignment and regulation methods are also discussed in the World Wireless Research Forum (WWRF). A white paper on spectrum for 4G includes not only the future spectrum allocation overview but also novel methods and technologies to use the 4G spectrum efficiently.

2.8 Conclusions

In this chapter we have approached 4G from a user-centric perspective. We started this journey discussing and analyzing the user, particularly his/her needs and expectations. A framework aimed at developing new services, taking as a starting point the user, was developed. As an intermediate step to new service development, we focused first on scenario development. Then, various services for each identified scenario were proposed and analysed. The current huge subscriber base and the forecasts that show further dramatic growth can be seen as an optimistic initial condition that will pave the way to a sound foundation for 4G. This is a key prerequisite but it does not guarantee the success of 4G. In our view this success depends first on the ability of service providers to develop useful, appealing, scalable (to different terminals), and flexible services. It also depends on the operator pricing policies. Fair fees, simple-to-calculate prices, and attractively low fares seem to be conditions that will have a strong influence on the initial acceptance and further development of 4G services. This is true in particular when 4G systems are thought to exploit a plethora of advanced multimedia-based services on a variety of possible terminals, with different types of service qualities involved (e.g., depending on terminal size and/or user selection), and from different providers. As important as the above items is the role of technology and how it would support the services. We think that 4G is highly justified from a user-centric view, as long as the aforementioned conditions are met. Technology-wise, we consider the open 4G vision of ITU as a suitable model for approaching and further developing 4G. Its importance lies in the fact that it attempts to integrate several developments into one common platform. These developments correspond to present and future realizations of several networks based on different access techniques. This concurrent approach gives equal opportunities to the main parties involved (namely telecom and IT manufacturers) for developing 4G technologies, particularly WWAN, WMAN, and WLAN access solutions. We saw that these three developing paths of constituent technologies are targeting 4G, and eventually each of these approaches will be an integral part of 4G. These solutions can be regarded as complementary and will altogether cover the whole range of radio scenarios. At the time 4G is launched, each of the mentioned developments will encompass legacy systems, enhanced versions of these systems, and possibly other solutions based on new technologies. It is expected that some overlapping of the capabilities of the constituent technology will occur, and hence, in addition to coexisting, the different access schemes will also compete. The vision of 4G as a monolithic, ubiquitous network of heterogeneous networks supporting a wide range of terminals with different capabilities and a variety of eclectic services is truly fascinating. Realizing this vision will require an enormous amount of effort from the involved parties, not only to develop the 4G component technologies but also to guarantee transparent and seamless operation in and between any possible scenarios, with any type of terminal, and exploiting any kind of available service.

References

[1]   Frattasi, S., et al., "A Pragmatic Methodology to Design 4G: From the User to the Technology," Springer: Lecture Notes in Computer Science, 3420, pp. 366-373.

[2]   Frattasi, S., et al., "A New Pragmatic Methodology to Design 4G," International Conference on Networking (ICN), Reunion Island, 2005, pp. 366–373.

[3]   ITU-R M.1645, "Framework and Overall Objectives of the Future Development of IMT-2000 and Systems Beyond IMT-2000."

[4]   Flower, Joe, "Spinning the Future," available at http://www.well.com/user/bbear/change12.html.

[5]   Linstone, H. A., and M. Turoff, The Delphi Method: Techniques and Applications, Reading, MA: Addison-Wesley, 1975.

[6]   Karlson, B., et al., Wireless Foresight: Scenarios of the Mobile World in 2015, New York: Wiley, 2003.

[7]   Chuang, J. C., "The Effects of Time Delay Spread on Portable Radio Communications Channels with Digital Modulation," IEEE Journal on Selected Areas in Communications, Vol. SAC-5, No. 5, June 1987, pp. 879-889.

[8]   Legutko, C., "Views on the Technical Radio Aspects Concerning Suitable Frequency Ranges for Systems Beyond IMT-2000," SIG 1 on the 12th WWRF meeting, Nov. 2004.

[9]   Oda, Y., et al., "Measured Path Loss and Multipath Propagation Characteristics in UHF and Microwave Frequency Bands for Urban Mobile Communications," VTC 2001 Spring, IEEE VTS 53rd, Vol. 1, May 6-9, 2001.

[10] Kitao, K., and S. Ichitsubo, "Path Loss Prediction Formula for Microcell in 400 MHz to 8 GHz Band," Electronics Letters, Vol. 40, No. 11, May 2004.

[11] Bertrand, G., A. Michalski, and L. R. Pench, Scenarios Europe 2010, European Commission, Forward Studies Unit, 1999.

[12] Flament, M., et al., "Telecom Scenarios for the 4th Generation Wireless Infrastructures," PCC Workshop '98, Stockholm, November 1998, pp. 11–15.

[13] Flament, M., et al., "Telecom Scenarios 2010," PCC Workshop '98, Stockholm, November 1998.

[14] Schwartz, P., The Art of the Long View, New York: Doubleday, 1996.

[15] mITF, Flying Carpet, Version 2.00, 2004.

[16] WWRF, Book of Vision 2001, 2001.

[17] 8F/328, "Chairman's Report of the 14th WP8F Meeting at Shanghai."

[18] Nobles, N., and F. Halsall, "Delay Spread Measurements Within a Building at 2 GHz, 5 GHz, and 17 GHz," IEE Colloquium on Propagation Aspects of Future Mobile Systems, October 25, 1996.

[19] Kim, J., "A Framework for Scenario/Service Development and its Application to 4G," The 12th WWRF, WG1, 2004.

[20] Katz, M., and F. Fitzek, "On the Definition of the Fourth Generation Wireless Communications Networks: The Challenges Ahead," International Workshop on Convergent Technologies (IWCT) 2005, Oulu, Finland, June 6-9, 2005.

3

Multiple Access Techniques

3.1 Introduction

After the introduction to 4G in the previous chapters, access techniques is the next natural topic to be discussed because the first and foremost challenge has always been to select a suitable access technique for defining and developing any mobile communications generation. At the moment we are struggling to find a solution to this challenge for 4G [1–6].

Figure 3.1 shows the history and evolution of access techniques over the generations of mobile communications systems. In the discussion about 2G systems in 1980s, there were two candidates for the radio access technique: the time division multiple access (TDMA) and code division multiple access (CDMA) schemes. The TDMA scheme was eventually adopted as the standard. In the discussion of 3G systems in 1990s, there were also two candidates: the CDMA scheme that was not adopted in the older systems and the orthogonal frequency division multiplexing (OFDM)-based multiple access scheme called band division multiple access (BDMA) [7]. CDMA was adopted as the standard. If history is repeated—that is, if the access technique that was once not adopted can become a standard in a new generation system—OFDM-based technique looks to be promising as a 4G standard.

The following list provides some of our justifications:

1. Multicarrier techniques can combat hostile frequency selective fading encountered in mobile communications. The robustness against frequency selective fading is very attractive especially for high-speed data transmission [8].

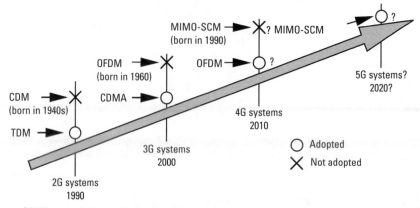

CDMA: code division multiple access
TDMA: time division multiple access
OFDM: orthognal frequency division multiplexing
MIMO-SCM: multiple input multiple output–single-carrier modulation

Figure 3.1 History of mobile communications generations in terms of adopted access technique.

2. The OFDM scheme has been well matured through research and development for high rate wireless LANs and terrestrial digital video broadcasting. A great deal of know-how has been developed on OFDM.

3. By combining OFDM with CDMA, we can have synergistic effects, such as enhancement of robustness against frequency selective fading and high scalability in possible data transmission rate.

Figure 3.2 shows the advantages of multicarrier techniques.

This chapter is organized as follows. Section 3.2 presents the definition and the classification of multiple access protocols. Sections 3.3, 3.4, 3.5, and 3.6 introduce, respectively, multicarrier code division multiple access (MC-CDMA), orthogonal frequency division multiple access (OFDMA), orthogonal frequency division multiplexing code division multiple access slow frequency hopping (OFDM-CDMA/SFH), and variable spreading factor–orthogonal frequency and code division multiplexing (VSF-OFCDM). Section 3.7 concludes the chapter.

3.2 Multiple Access Protocols

Agreement among users on the means of communication is known as protocol. When users use a common medium for communications, it is called multiple

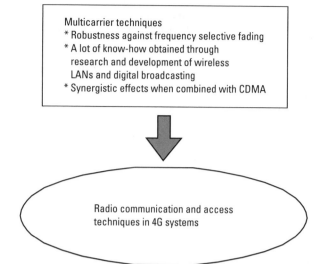

Figure 3.2 Advantages of multicarrier techniques for 4G systems.

access. Thus, the multiple access protocol is defined as the agreement and set of rules among users for successful transmission of information using a common medium. Whenever some resource is used and thus accessed by more than one independent user, the need for a multiple access protocol arises. In the absence of such a protocol, conflicts occur if more than one user tries to access the resource at the same time. Therefore, the multiple access protocol should avoid or at least resolve these conflicts. Thus, a *multiple access technique is defined as a function sharing a (limited) common transmission resource among (distributed) terminals in a network.*

The multiple access protocols addressed in this chapter are those used in communication systems in which the resource to be shared is the communication channel. In this case, the reason for sharing the resource is mainly the connectivity environment. In wireless communication systems, an added reason is the scarceness of radio resource; there is only one ether [1]. When, instead of a common medium with an access for every user, there is a network that consists of point-to-point links (different medium), but one of these links is simultaneously shared by many users, the multiplexing technique is needed. Multiplexing is the transmission of different information in the same physical link. Table 3.1 shows the difference between multiple access and multiplexing.

The design of a protocol is usually accomplished with a specific goal (environment) in mind, and the properties of the protocol are mainly determined by the design goal. But if we rule out the environment-specific properties, we can still address a number of properties that any good multiple access protocol should possess:

Table 3.1

Difference Between Multiple Access and Multiplexing

	Multiple Access	**Multiplexing**
Resource	Network	Link
Terminal connectivity	Matrix	Point-to-(multi)point
Switching	Yes	No
	(≤ 3)	(OSI ≤ 2)
Possible	(OSI: open systems interconnection)	
Topologies	Bus Star Ring Tree	Path(s)
Control	Central Distributed	Terminal

- The first and foremost task of the multiple access protocols addressed here is to share the common transmission channel among the users in the system. To do this, the protocol must control the way in which the users access (transmit onto) the channel by requiring that the users conform to certain rules. The protocol controls the allocation of channel capacity to the users.

- It should perform the allocation such that the transmission medium is used efficiently. Efficiency is usually measured in terms of channel throughput and the delay of the transmissions.

- The allocation should be fair toward individual users; that is, it should not take into account any priorities that might be assigned to the users—each user should (on the average) receive the same allocated capacity.

- The protocol should be flexible in allowing different types of traffic (e.g., voice and data).

- It should be stable. This means that if the system is in equilibrium, an increase in load should move the system to a new equilibrium point. With an unstable protocol an increase in load will force the system to continue to drift to even higher load and lower throughput.

- The protocol should be robust with respect to equipment failure and changing conditions. If one user does not operate correctly, this should affect the performance of the rest of the system as little as possible.

In this chapter, the wireless mobile environment is what we are most interested in. In such an environment, we can be more specific about some of the protocol properties, especially on the robustness with respect to changing conditions. In the wireless mobile environment, the protocol should be able to deal with:

- The hidden terminal problem: two terminals are out of range (hidden from) of each other by a hill, a building, or some physical obstacle opaque to UHF signals, but both are within the range of the central or base station;

- The near-far effect: transmissions from distant users are more attenuated than transmissions from users close by;

- The effects of multipath fading and shadowing experienced in radio channels;

- The effects of cochannel interference in cellular wireless systems caused by the use of the same frequency band in different cells.

Many of the protocol properties mentioned above are conflicting, and a trade-off has to be made during the protocol design. The trade-off depends on the environment and the specific use for the protocol one has in mind.

3.2.1 Classification of Multiple Access Protocol

Starting in 1970 with the ALOHA protocol, a number of multiple access protocols have been developed. Numerous ways have been suggested to divide these protocols into groups [1, 2, 9, 10]. In this chapter, multiple access protocols are classified into three main groups (see Figure 3.3): the contentionless protocols, the contention protocols, and the hanging protocols.

The contentionless (or scheduling) protocols avoid the situation in which two or more users access the channel at the same time by scheduling the transmissions of the users. This is either done in a fixed fashion where each user is allocated part of the transmission capacity, or in a demand-assigned fashion where the scheduling only takes place between the users that have something to transmit.

With the contention (or random access) protocols, a user cannot be sure that a transmission will not collide because other users may be transmitting (accessing the channel) at the same time. Therefore, these protocols need to resolve conflicts if they occur.

The contention protocols are further subdivided into repeated random protocols and random protocols with reservation. In the latter, the initial transmission of a user uses a random access method to get access to the channel. However, once the user accesses the channel, further transmissions of that user

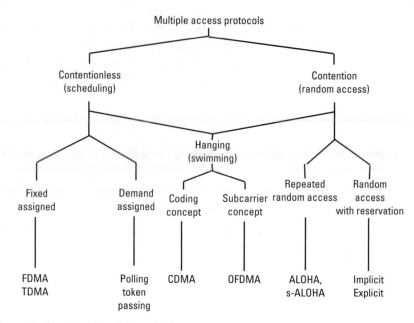

Figure 3.3 Classification of the multiple access protocols.

are scheduled until the user has nothing more to transmit. Two major types of protocols are known as implicit and explicit reservations. Explicit reservation protocols use a short reservation packet to request transmission at scheduled times. Implicit reservation protocols are designed without the use of any reservation packet.

The hanging (or swimming) protocols do not belong to either the contentionless or the contention protocols. They fall between the two groups. In principle, it is a contentionless protocol where a number of users are allowed to transmit simultaneously without conflict. However, if the number of simultaneously transmitting users rises above a threshold, contention occurs.

The hanging protocols are subdivided into techniques based on coding and subcarrier concepts. The hanging protocols are achieved either by assigning unique code to each user or by providing each user with a fraction of the available number of subcarriers. With hanging protocols forming a separate group and taking into account the subgroups of the contentionless and contention protocols, we end up with six categories.

3.2.1.1 Contentionless (Scheduling) Multiple Access Protocols

The contentionless multiple access protocols avoid the situation in which multiple users try to access the same channel at the same time by scheduling the transmissions of all users. The users transmit in an orderly scheduled manner so every transmission will be a successful one. The scheduling can take two forms:

1. *Fixed assignment scheduling.* With these types of protocols, the available channel capacity is divided among the users such that each user is allocated a fixed part of the capacity, independent of its activity. The division is done in time or frequency. The time division results in the TDMA protocol, where transmission time is divided into frames and each user is assigned a fixed part of each frame, not overlapping with parts assigned to other users. TDMA is illustrated in Figure 3.4(a). The frequency division results in the frequency division multiple access (FDMA) protocol, where the channel bandwidth is divided into nonoverlapping frequency bands and each user is assigned a fixed band. Figure 3.4(b) illustrates FDMA.

2. *Demand assignment scheduling.* A user is only allowed to transmit if he/she is active (if he/she has something to transmit). Thus, the *active* (or ready) users transmit in an orderly scheduled manner. Within the demand assignment scheduling, we distinguish between centralized control and distributed control. With centralized control, a single entity schedules the transmissions. An example of such a protocol is the roll-call polling protocol. With distributed control, all users are involved in the scheduling process and such a protocol is the token-passing protocol.

3.2.1.2   Contention (Random) Multiple Access Protocols

With contention multiple access protocols there is no scheduling of transmissions. This means that a user getting ready to transmit does not have exact knowledge of when it can transmit without interfering with the transmissions of

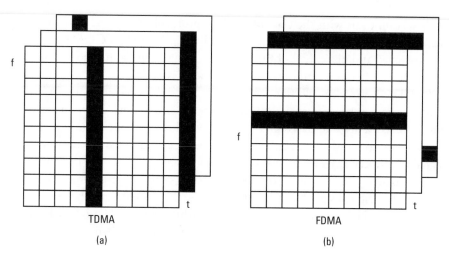

TDMA

(a)

FDMA

(b)

Figure 3.4   (a) TDMA and (b) FDMA.

other users. The user may or may not know of any ongoing transmissions (by sensing the channel), but it has no exact knowledge about other ready users. Thus, if several ready users start their transmissions more or less at the same time, all of the transmissions may fail. This possible transmission failure makes the occurrence of a successful transmission a more or less random process. The random access protocol should resolve the contention that occurs when several users transmit simultaneously.

We subdivide contention multiple access protocols into two groups: the repeated random access protocols [e.g., pure (p)-ALOHA, slotted (s)-ALOHA, carrier sense multiple access (CSMA), ISMA), and random access protocols with reservation [e.g., reservation (r)-ALOHA, packet reservation multiple access (PRMA)]. With the former protocols, every transmission a user makes is as described above. With every transmission there is a possibility of contention. With the latter protocols, only in its first transmission does a user not know how to avoid collisions with other users. However, once a user has successfully completed its first transmission (once the user has access to the channel), future transmissions of that user will be scheduled in an orderly fashion so that no contention can occur. Thus, after a successful transmission, part of the channel capacity is allocated to the user, and other users will refrain from using that capacity. The user loses its allocated capacity if, for some time, it has nothing to transmit.

3.2.1.3   Hanging (Swimming) Multiple Access Protocols

The hanging multiple access protocols are subdivided into coding-based protocols (CDMA protocols) and subcarrier-based protocols (OFDMA).

CDMA Protocols

This section deals with the class of CDMA protocols, which rely on coding to achieve their multiple access property. The basic principles of CDMA protocols are discussed and placement of the protocols between the contentionless (scheduling) and contention (random access) protocols is explained.

CDMA protocols do not achieve their multiple access property by a division of the transmissions of different users in either time or frequency, but instead make a division by assigning each user a different code (Figure 3.5). This code is used to transform a user's signal into a wideband signal (spread-spectrum signal). If a receiver receives multiple wideband signals, it uses the code assigned to a particular user to transform the wideband signal received from that user back to the original signal. During this process, the desired signal power is compressed into the original signal bandwidth, while the wideband signals of the other users remain wideband signals and appear as noise when compared with the desired signal.

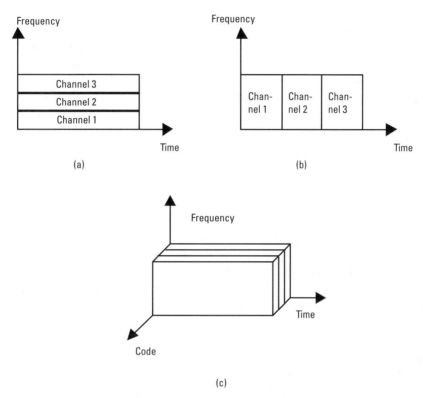

Figure 3.5 Multiple access schemes: (a) FDMA; (b) TDMA; and (c) CDMA.

As long as the number of interfering users is not too large, the signal-to-noise ratio (SNR) is large enough to extract the desired signal without error. Thus, the protocol behaves as a contentionless protocol. However, if the number of users rises above a certain limit, the interference becomes too large for the desired signal to be extracted and contention occurs, making the protocol interference limited. Therefore, the protocol is basically contentionless unless too many users access the channel at the same time. This is why we place CDMA protocols between contentionless and contention protocols.

There are several ways to classify CDMA schemes. The most common is the division based on the modulation method used to obtain the wideband signal. This division leads to three types of CDMA: direct sequence (DS), frequency hopping (FH), and time hopping (TH), as illustrated in Figure 3.6[1]. In DS-CDMA, spectrum is spread by multiplying the information signal with a pseudo-noise sequence, resulting in a wideband signal. In the frequency hopping spread spectrum, a pseudo-noise sequence defines the instantaneous transmission frequency. The bandwidth at each moment is small, but the total bandwidth over, say, a symbol period is large. Frequency hopping can either be fast (several hops over one symbol) or slow (several symbols transmitted during one hop). In the time hopping spread spectrum, a pseudo-noise sequence defines the transmission moment. Furthermore, combinations of these techniques are possible. In this chapter we focus on DS-CDMA because it is the technique used for third generation wideband CDMA proposals. Wideband CDMA is defined as a direct sequence spread spectrum multiple access scheme where the information is spread over a bandwidth of approximately 5 MHz or more.

OFDMA Protocols

OFDM, a special form of multicarrier modulation, isused as multiple access schemes in wireless communications. In OFDM, densely spaced subcarriers with overlapping spectra are generated using fast Fourier transform (FFT) [3]. Signal waveforms are selected in such a way that the subcarriers maintain their orthogonality despite the spectral overlap. Usually, CDMA or TDMA is used with OFDM to achieve multiple access capability.

Basically, orthogonal frequency resources (i.e., orthogonal subcarriers) can be shared among users, which is the simplest multiple access technique based on OFDM modulation. Besides, there are a number of hybrid multiple access techniques that can be found in literature. Here, hybrid means an amalgamation of OFDM and multiple access techniques (with the main accent to the spread spectrum) to provide an efficient multiuser scenario with very high data rate. The following techniques will be discussed in this chapter [1–38]:

1. MC-CDMA [3, 5];
2. OFDMA [3];
3. OFDM-CDMA-SFH [6];
4. VSF-OFCDMA [37, 38].

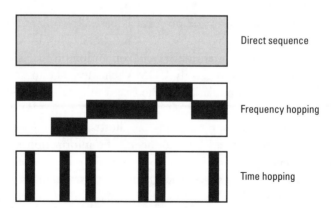

Direct sequence

Frequency hopping

Time hopping

Figure 3.6 Direct sequence, frequency hopping, and time hopping CDMA.

3.3 MC-CDMA System

The OFDM scheme is insensitive to frequency selective fading, but it has severe disadvantages such as difficulty in subcarrier synchronization and sensitivity to frequency offset and nonlinear amplification; on the other hand, the CDMA scheme has robustness against frequency selective fading. Therefore, any synergistic effect might not be expected in combining an OFDM scheme with a CDMA scheme. However, the combination has two major advantages. One is its own capability to lower the symbol rate in each subcarrier enough to have a quasi-synchronous signal reception in uplink. The other is that it can effectively combine the energy of the received signal scattered in the frequency domain. Especially for high-speed transmission cases where a DS-CDMA receiver could see 20 paths in the instantaneous impulse response, a 20-finger Rake combiner would be impossible to implement for the DS-CDMA receiver, whereas an MC-CDMA receiver would be possible although it would lose the energy of the received signal in the guard interval.

The MC-CDMA transmitter spreads the original signal using a given spreading code in the frequency domain. In other words, a fraction of the symbol corresponding to a chip of the spreading code is transmitted through a different subcarrier. For multicarrier transmission, it is essential to have frequency nonselective fading over each subcarrier. Therefore, if the original symbol rate is high enough to become subject to frequency-selective fading, the signal needs to be serial-to-parallel converted first before being spread over the frequency domain. The basic transmitter structure of MC-CDMA scheme is similar to that of a normal OFDM scheme. The main difference is that the MC-CDMA scheme transmits the same symbol in parallel through many subcarriers, whereas the OFDM scheme transmits different symbols.

Figure 3.7(a) shows the MC-CDMA transmitter P parallel data sequences $(a_{j,0}(i), a_{j,1}(i), \ldots, a_{j,P-1}(i))$ and then each serial/parallel converter output is multiplied with the spreading code with length K_{MC}. All the data in total $N = P \times K_{MC}$ (corresponding to the total number of subcarriers) are modulated in baseband by the inverse discrete Fourier transform (IDFT) and converted back into serial data. The guard interval Δ is inserted between symbols to avoid intersymbol interference caused by multipath fading, and finally the signal is transmitted after RF up-conversion. The complex equivalent low-pass transmitted signal is written as

$$S_{MC}^{j}(t) = \sum_{i=-\infty}^{+\infty} \sum_{p=0}^{P-1} \sum_{m=0}^{K_{MC}-1} a_{j,p}(i) d_j(m) p_s(t - iT_s') e^{j2\pi(Pm+p)\Delta f'(t-T_s')} \qquad (3.1)$$

$$T_s' = PT_s \qquad (3.2)$$

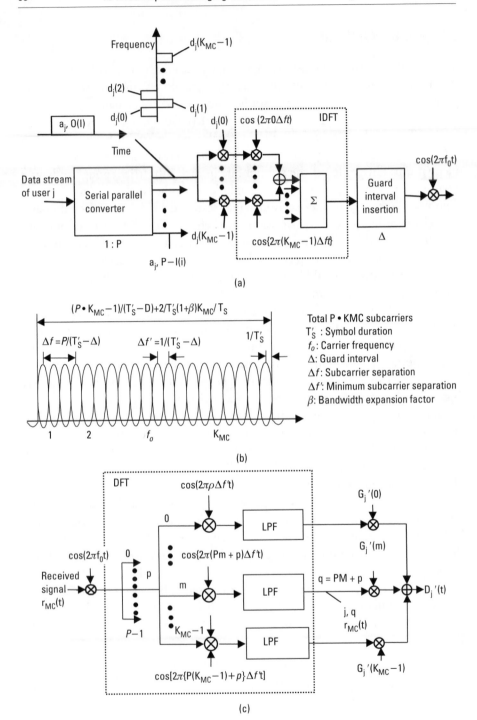

Figure 3.7 MC-CDMA system: (a) transmitter; (b) power spectrum of transmitted signal; and (c) receiver.

$$\Delta f' = 1/(T_s' - \Delta) \tag{3.3}$$

where $\{d_j(0), d_j(1), ..., d_j(K_{MC}-1)\}$ is the spreading code with length K_{MC}, T_s' is the symbol duration at subcarrier, $\Delta f'$ is the minimum subcarrier separation, and $p_s(t)$ is the rectangular symbol pulse waveform defined as

$$p_s(t) = \begin{cases} 1 & (-\Delta \leq t \leq T_s' - \Delta) \\ 0 & (\text{otherwise}). \end{cases} \tag{3.4}$$

The bandwidth of the transmitted signal spectrum is written as [see Figure 3.7(b)]

$$B_{MC} = (P \cdot K_{MC} - 1)/(T_s' - \Delta) + 2/T_s' \tag{3.5}$$

$$\begin{aligned} &\approx K_{MC}/T_s/(1 - \Delta/P) \\ &= (1+\beta)K_{MC}/T_s \\ &\beta = \Delta/P \qquad (0 \leq \beta \leq 1.0) \end{aligned} \tag{3.6}$$

where β is the bandwidth expansion factor associated with the guard interval insertion.

Note that, in (3.1), no spreading operation is done in the time domain. Equation (3.2) shows that the symbol duration at subcarrier level is P times as long as the original symbol duration because of serial/parallel conversion. Although the minimum subcarrier separation is given by (3.3), the subcarrier separation for $a_{j,p}(i)$ is $\Delta f = P/(T_s' - \Delta)$ [see the hatched subcarrier power spectra in Figure 3.7(b)]. Therefore, when setting K_{MC} to 1, the transmitted waveform given by (3.1) becomes all the same as an OFDM waveform with P subcarriers.

On the other hand, the received signal is written as

$$r_{MC}(t) = \sum_{j=1}^{J} \int_{-\infty}^{+\infty} S_{MC}^j(t-\tau) \otimes h^j(\tau;t)d\tau + n(t)$$

$$= \sum_{j=-\infty}^{+\infty} \sum_{p=0}^{P-1} \sum_{m=0}^{K_{MC}-1} \sum_{j=1}^{J} z_{m,p}^j(t) a_{j,p}(i) d_m^j p_s(t - iT_s') e^{j2\pi(Pm+p)\Delta f't} + n(t) \tag{3.7}$$

where $z_{m,p}^j(t)$ is the received complex envelope at the $(mP+p)^{\text{th}}$ subcarrier of the jth user.

The MC-CDMA receiver requires coherent detection for successful despreading operation. Figure 3.7(c) shows the MC-CDMA receiver for the jth user. After down-conversion, the m-subcarrier components ($m = 0, 1, ..., K_{MC}-1$) corresponding to the received data $a_{j,p}(i)$ is first coherently detected with DFT and then multiplied with the gain $G_j(m)$ to combine the energy of the received signal scattered in the frequency domain. The decision variable is the sum of the weighted baseband components given by (we can omit the subscription p without loss of generality)

$$D_{MC}^{j'}(t = iT_s) = \sum_{m=0}^{K_{MC}-1} G_{j'}(m)y(m) \qquad (3.8)$$

$$y(m) = \sum_{j=1}^{J} z_m^j(iT_s)a_j d_m^j + n_m(iT_s) \qquad (3.9)$$

where $y(m)$ and $n_m iTs$ are the complex baseband component of the received signal after down-conversion and the complex additive Gaussian noise at the mth subcarrier at $t = iTs$, respectively. The system adapts different combining strategies to recover the signal. They are orthogonality restoring combining (ORM), equal gain combining (EGC), maximal ratio combining (MRC), and minimum mean square error combining (MMSEC). The techniques are discussed in [3, 5, 6].

3.3.1 MC-CDMA System Design

To determine the number of subcarriers and the length of guard interval, we derive the autocorrelation function of the received signal. The received signal for the jth user is given by

$$r_{MC}^j(t) = \int_{-\infty}^{+\infty} s_{MC}^j(t-\tau) \otimes h_{jg}(\tau;t)d\tau + n(t) \qquad (3.10)$$

where $h^j(\tau;t)$ is the complex equivalent low-pass time-variant impulse response with L received paths and $\{\tau_l\}$ is classified as follows:

$$\begin{aligned} 0 \leq \tau_l \leq \Delta \qquad & (l = 1,...,L_1) \\ \Delta \leq \tau_l \leq T_s \qquad & (l = L_1 + 1,..., L_1 + L_2 (= L)) \end{aligned} \qquad (3.11)$$

The Fourier coefficient of the 9th ($q = mP + p$) subcarrier at $t = iT_s'$ is given by [see Figure 3.7(c)]

$$r_{MC}^{j,q}\left(iT_s'\right)=\frac{1}{T_s'-\Delta}\int_{iT_s'}^{iT_s'+T_s'-\Delta}r_{MC}^{j}(t)e^{-j2\pi q\Delta f'(t-iT_s')}dt \qquad (3.12)$$

The normalized autocorrelation function of the qth subcarrier between $t=iT_s'$ and $t=(i-1)T_s'$ for the jth user is written as

$$R_{MC}^{j,q}\left(\Delta,N,R',f_D,\tau_{RMS}\right)=\frac{E\left[r_{MC}^{j,q}\left(iT_s'\right)\cdot r_{MC}^{*j,q}\left((i-1)T_s'\right)\right]}{E\left[r_{MC}^{j,q}\left(iT_s'\right)\cdot r_{MC}^{*j,q}\left(iT_s'\right)\right]}$$
$$=\frac{\sigma_{S1}^2}{\sigma_{S2}^2+\sigma_I^2+\sigma_n^2} \qquad (3.13)$$

where σ_{S1}^2, σ_{S2}^2, σ_I^2, and σ_n^2 are given in [3].

In an OFDM scheme, generally, when the transmission rate, R', (in the case of MC-CDMA scheme) is given, the transmission performance becomes more sensitive to time-selective fading as the number of subcarriers N increases, because the longer symbol duration means an increase in the amplitude and phase variation during a symbol, causing an increased level of inter-carrier interface (ICI). As N decreases, the modulation becomes more robust to fading in time, but it becomes more vulnerable to delay spread, as the ratio of delay spread and symbol duration increases (see Figure 3.8). The latter is not necessarily true if the guard time is kept at a fixed value, but as the symbol duration decreases, a fixed guard interval (Δ) means an increased loss of power (see Figure 3.9). Therefore, for given R (R), f_D, and τ_{RMS}, there exists an optimum that minimizes the bit error rate (BER) in both N and Δ [13].

In the MC-CDMA scheme, N_{opt} and Δ_{opt} maximize the autocorrelation function (ACF) given by (3.13), because it means a measure to show how much the received signal is distorted in the time-frequency, selective fading channel (i.e., how we can place the signal on the time-frequency plane so that it suffers from minimum distortion?):

$$\left[N_{opt},\Delta_{opt}\right]=\arg\left\{\max R_j\left(N,\Delta|R',f_D,\tau_{RMS}\right)\right\} \qquad (3.14)$$

Therefore, with (3.14), we determine two parameters, N_{opt} and Δ_{opt}.

3.3.2 Summary

Considering the results presented in [3, 5, 6] for the downlink channel, we can conclude that it could be difficult for a DS-CDMA receiver to employ all the received signal energy scattered in the time domain, whereas an MC-CDMA

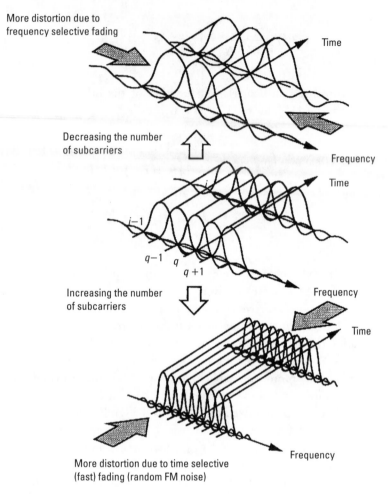

More distortion due to
frequency selective fading

Time

Decreasing the number
of subcarriers

Frequency

Time

i

$i-1$

$q-1$ q
    $q+1$

Increasing the number
of subcarriers

Frequency

Time

More distortion due to time selective
(fast) fading (random FM noise)

Frequency

Figure 3.8 Optimum in the number of subcarriers.

receiver can effectively combine all the received signal energy scattered in the frequency domain. A DS-CDMA receiver needs to make efforts to select larger paths; on the other hand, an MC-CDMA receiver does not care about where the received signal energy is. This is a significant advantage of the MC-CDMA scheme over a DS-CDMA scheme, and it makes the MMSEC-based MC-CDMA a promising access scheme in a downlink channel.

In uplink [3], the MMSEC performs well only for the single user case, otherwise it performs poorly. This is because the code orthogonality among users is totally distorted by the instantaneous frequency response. Therefore, in the uplink application, a multiuser detection scheme is required, which jointly detects the signals to mitigate the nonorthogonal properties [20]. There are many other detectors reported in the literature [21, 22].

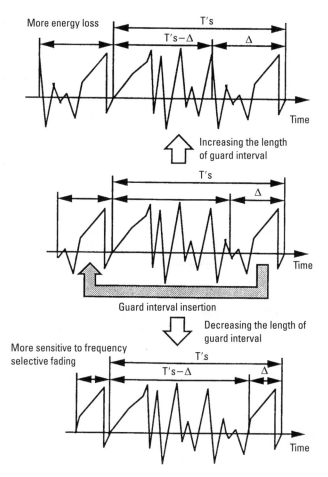

Figure 3.9 Optimum in the length of guard interval.

3.4 Orthogonal Frequency Division Multiple Access (OFDMA)

The previous section described some ways in which OFDM could be used both as a modulation scheme and as part of the multiple access technique, by applying a spreading code in the frequency domain. In this section, a variation on this theme is described, namely orthogonal frequency division multiple access. In OFDMA, multiple access is realized by providing each user with a fraction of the available number of subcarriers. In this way, it is equal to ordinary FDMA; however, OFDMA avoids the relatively large guard bands that are necessary in FDMA to separate different users. An example of an OFDMA time-frequency grid is shown in Figure 3.10, where seven users (a to g) each use a certain fraction—which may be different for each user—of the available subcarriers. This particular example in fact is a mixture of OFDMA and TDMA, because each

Frequency										
a		d		a		d		a		d
a		d		a		d		a		d
a	c	e		a	c	e		a	c	e
a	c	e		a	c	e		a	c	e
b		e	g	b		e	g	b		e	g
b		e	g	b		e	g	b		e	g
b		f	g	b		f	g	b		f	g
b		f	g	b		f	g	b		f	g

Time

Figure 3.10 Example of the time-frequency grid with seven OFDMA users (a to g), all of which have a fixed set of subcarriers every four time slots.

user only transmits in one out of every four time slots, which may contain one or several OFDM symbols.

3.4.1   Frequency Hopping OFDMA

In the previous example of OFDMA, every user had a fixed set of subcarriers. It is a relatively easy change to allow hopping of the subcarriers per time slot, as depicted in Figure 3.11. Allowing hopping with different hopping patterns for each user actually transforms the OFDMA system in a frequency hopping CDMA system. This has the benefit of increased frequency diversity, because each user uses all of the available bandwidth, as well as the interference averaging benefit that is common for all CDMA variants. By using forward error correction (FEC) coding over multiple hops, the system can correct for subcarriers in deep fades or subcarriers that are interfered by other users. Because the interference and fading characteristics change for every hop, the system performance depends on the average received signal power and interference, rather than on the worst case fading and interference power.

A major advantage of frequency hopping CDMA systems over direct-sequence or multicarrier CDMA systems is that it is relatively easy to eliminate intracell interference by using orthogonal hopping patterns within a cell. An example of such an orthogonal hopping set is depicted in Figure 3.12. For N subcarriers, it is always possible to construct N orthogonal hopping

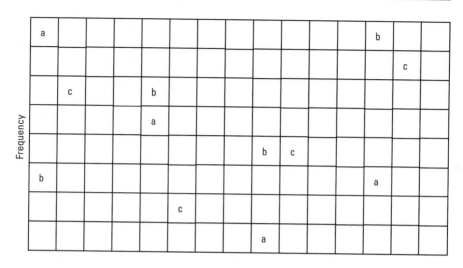

Time

Figure 3.11 Example of the time-frequency grid with three hopping users (a, b, and c), which all have one hop every four time slots.

patterns. Some useful construction rules for generating hopping patterns can be found in [23].

3.4.2 Differences Between OFDMA and MC-CDMA

The main difference between OFDMA and the MC-CDMA techniques is that users within the same cell use a distinct set of subcarriers, while in MC-CDMA, all users use all subcarriers simultaneously. To distinguish different users,

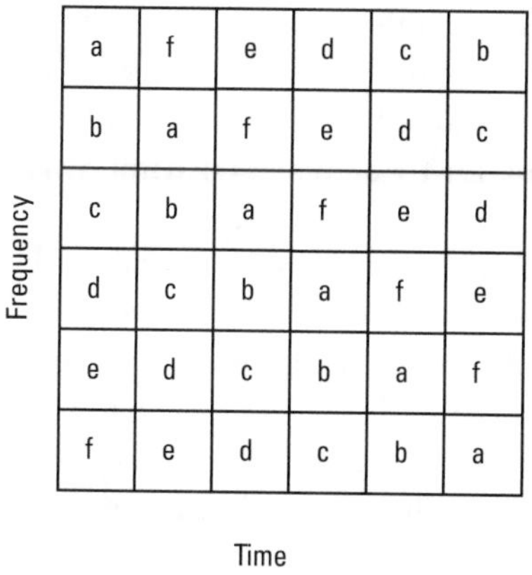

a	f	e	d	c	b
b	a	f	e	d	c
c	b	a	f	e	d
d	c	b	a	f	e
e	d	c	b	a	f
f	e	d	c	b	a

Frequency (vertical axis) / Time (horizontal axis)

Figure 3.12 Example of six orthogonal hopping patterns with six different hopping frequencies.

orthogonal or near-orthogonal spreading codes are used in MC-CDMA. Because of code distortion by multipath fading channels, however, MC-CDMA loses its orthogonality in the uplink even for a single cell. This makes rather complicated equalization techniques necessary, which introduce a loss in SNR performance and diminish the complexity advantage of OFDM over single-carrier techniques. OFDMA does not have this disadvantage, because in a single cell, all users have different subcarriers, thereby eliminating the possibility of intersymbol or intercarrier interference. Hence, OFDMA does not suffer from intracell interference, provided that the effects of frequency and timing offsets between users are kept at a sufficiently low level. This is a major advantage of OFDMA compared with MC-CDMA and DS-CDMA, because in those systems, intracell interference is the main source of interference. A typical ratio of intracell and intercell interference is 0.55 [24]. Because the system capacity is inversely proportional to the total amount of interference power, a capacity gain of 2.8 can be achieved by eliminating all intracell interference, so that is the maximum capacity gain of OFDMA over DS-CDMA and MC-CDMA networks.

The main advantages of using CDMA in general or MC-CDMA in particular is interference averaging. In CDMA, the interference consists of a much larger number of interfering signals than the interference in a non-CDMA system. Each interfering signal is subject to independent fading, caused by shadowing and multipath effects. For both the CDMA and the non-CDMA system, an outage occurs when the total interference power (after despreading for the CDMA system) exceeds some maximum value. In a non-CDMA system, the interference usually consists of a single or a few cochannel interferers. Because of fading, the interference power is fluctuating over a large range, so a large fading margin has to be taken into account, which reduces system capacity. In a CDMA system, the interference is a sum of a large number of interfering signals. Because all these signals fade independently, the fluctuation in the total interference power is much less than the power fluctuation of a single interfering signal. Hence, in a CDMA system the fading margin can be significantly smaller than the margin for a non-CDMA system. This improvement in margin largely determines the capacity gain of a CDMA system.

In OFDMA, interference averaging is obtained by having different hopping patterns within each cell. The hopping sequences are constructed in such a way that two users in different cells interfere with each other only during a small fraction of all hops. In a heavily loaded system, many hops will interfere, but the interference will be different for each hop. Hence, by FEC across several hops, the OFDMA performance will be limited by the average amount of interference rather than the worst-case interference. An additional advantage of OFDMA over DS-CDMA and MC-CDMA is that there are some relatively simple ways to reduce the amount of intercell interference. For instance, the receiver can estimate the signal quality of each hop and use this information to give heavily interfered hops a lower weight in the decoding process.

Another important feature of CDMA is the possibility of performing soft handover by transmitting two signals from different base stations simultaneously on the same channel to one mobile terminal. Combining the signals from different base stations gives a diversity gain that significantly reduces the fading margin, because the probability that two base stations are in a fade is much smaller than the probability that one base station is in a fade. Less fading means that less power has to be transmitted, and hence less interference is generated, which gives an improvement in the capacity of the system. A nice feature of CDMA soft handover is that it has no impact on the complexity of the mobile terminal; as far as the mobile terminal is concerned, the overlapping signals of different base stations have the same effect as overlapping signals caused by multipath propagation.

For OFDMA systems, two basic soft handover methods exist, applicable to both the uplink and the downlink: base station-to-mobile and mobile-to-base station). A requirement for both methods is that the transmissions from and to the base stations are synchronized such that the delay differences at the two base stations are well within the guard time of the OFDM symbols.

The first technique is to use the same set of subcarriers and the same hopping sequence in two cells to connect to two base stations. Hence, in the downlink the mobile receives a sum of two signals with identical data content. The mobile is not able to distinguish between the two base stations; the effect of soft handover is similar to that of adding extra multipath components, increasing the diversity gain. This type of soft handover is similar to soft handover in DS-CDMA networks.

The second way for soft handoff is to use different sets of subcarriers in two cells. In contrast to the first method, in the downlink the mobile has to distinguish now between the two base stations. It has to demodulate the signals from the two base stations separately, after which they can be combined, preferably by using maximal ratio combining. This type of soft handover is similar to the one that could be used in a non-CDMA network.

Advantages of the second method over the first—in the downlink—are an increased SNR gain because of receiver diversity, and more freedom for the base stations to allocate available subcarriers. In the first method, base stations are forced to use the same subcarriers. A main advantage of the first method is its simpler implementation: no additional hardware is needed, only some extra protocol features to connect to two base stations simultaneously. The second method does require extra hardware, because it has to demodulate an extra set of subcarriers. Furthermore, it has to perform extra processing for the maximal ratio combining of the signals from the different base stations.

3.4.3   Discussions

OFDMA is a very flexible multiple access scheme. In combination with frequency hopping, it offers all the benefits of direct-sequence CDMA or multicarrier CDMA systems, with the following additional advantages:

- OFDMA can achieve a larger system capacity because it is not affected by intracell interference, which is the dominant source of interference in DS-CDMA and MC-CDMA.

- OFDMA is flexible in terms of the required bandwidth; it can easily be scaled to fit in a certain available piece of spectrum simply by changing the number of used subcarriers. DS-CDMA and MC-CDMA require a fixed and relatively large chunk of spectrum because of the fixed spread-spectrum chip rate.

- OFDMA seems more suited than DS-CDMA and MC-CDMA for support of large data rates. When the data rates per user become larger, the spreading gain becomes lower, so the performance of a CDMA system converges to that of a nonspread system. In that case, OFDMA with dynamic channel allocation instead of frequency hopping makes more sense than multicode DS-CDMA or MC-CDMA.

3.5   OFDM-CDMA-SFH (Hybrid)

This is a new approach [6, 29–35]. In this case (Figure 3.13), suppose we have an 8-bit CDMA signal ("word"). We feed this word to a serial/parallel (S/P) converter. After this step comes a crucial difference. We do not carry out any Walsh spreading, but we straightaway carry out an 8-point inverse fast Fourier transform (IFFT) (OFDM modulation). This means that one OFDM symbol equals the entire 8-bit CDMA "word" that we need to transmit. Hence, our information rate is eight times faster than in the previous two procedures. At the receiver, we carry out an FFT, but with another crucial difference in that we do not use RAKE combiners. There is no need in our case to use RAKE combiners, since we are not carrying out Walsh spreading in the frequency domain, in the interests of frequency diversity. On the contrary, we will accept the risk that some subcarriers will be in deep fade and correct for this eventuality using FEC coding (coded OFDM) and/or interleaving. This is different from the OFDM-CDMA approach, wherein RAKE combiners are used after OFDM demodulation in order to take advantage of the entire frequency spread of that particular bit. In reality, if we use a N-point OFDM system, we will need to use a N finger RAKE combiner. This is extremely costly, and a compromise can be achieved by using a lesser number of fingers (e.g., a seven finger combiner. This

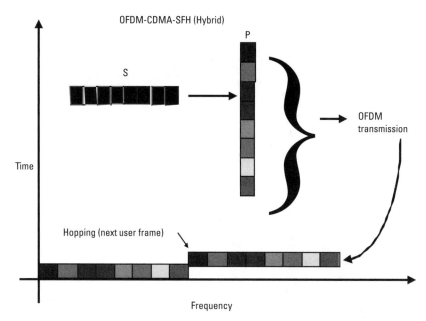

Figure 3.13 OFDM-CDMA-SFH (Hybrid) approach.

means that we do not take advantage of the entire frequency spread anyway. In a way, this is a waste of resources. Therefore, after FFT we carry out a parallel/serial (P/S) conversion and then despread the CDMA signal. For user separation, we can use Walsh spreading in the transmitter. But this is done before the S/P converter in the transmitter (i.e., in the time domain).

In the receiver we despread the Walsh signal after the P/S converter. This will ensure the user separation. There are other spin-offs from this technique and these are discussed later in this chapter. Finally, in the frequency-time diagram, we have shown slow frequency hopping of the entire symbol.

This proposal pertains to the downlink as well as to the uplink, the only difference being that for the synchronous downlink we can apply orthogonal Walsh-Hadamard sequences leading to the well-known user separation for MC-CDMA systems. On the other hand, in the asynchronous uplink scenario, pseudo-noise-sequences are used with the drawback of high multiple access interference (MAI).

The overall concept discussed so far is shown in the schematic in Figure 3.14.

3.5.1 Description

It can be seen in the schematic that the transmitter and receiver are each divided into three sections:

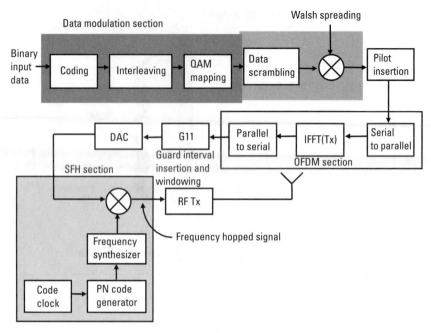

Hybrid transmitter

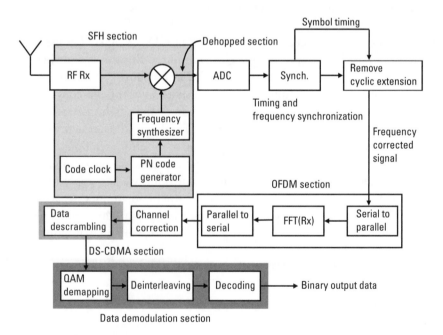

Hybrid receiver

Figure 3.14 Overall sytem schematic.

- Data modulation (demodulation) section;
- DS-CDMA section;
- SFH section.

The binary input data enters the data modulation section, where it is encoded by a FEC code. We can also use concatenated coding comprising a convolutional coding as an inner code followed by an outer coding as a block code (e.g., a Reed-Solomon code) [3]. This makes for a large coding gain with less implementation complexity as compared to a single code. This coding is followed by interleaving to randomize the occurrence of bit errors due to deep fades across certain subcarriers. This is followed by QAM mapping. Thereafter, the data enters the DS-CDMA section. In this section, the data is subjected to scrambling based on pseudonoise sequences. This aspect is similar to the implementation for the IS-95 system. Thereafter, Walsh coding is used to provide orthogonal covering, because pseudonoise sequences by themselves are insufficient to ensure user separation. The Walsh function matrix will be a $N \times N$ matrix, where N is the number of OFDM points. We then obtain what we can call a DS-CDMA signal. We insert pilot symbols after this step. In doing so, we must take care as regards the size of the Walsh matrix. For example, if our OFDM system has 32 points, we can have at most a 24 length Walsh sequence. This leaves eight subcarriers for pilots. This is assuming that we utilize all the subcarriers. In practice, this is not possible, because we need to leave the edge subcarriers unutilized. This problem is explained as follows. We need to allow for the skirt of the low-pass antialiasing filter in the receiver. Subcarriers that lie beyond the bandwidth of interest should contain zero information. This is because it is these subcarriers that will lie along the slope of the low pass filter. If they contained information then their amplitudes will not be uniform since they lie along the slope. It will be recalled that the fundamental assumption for orthogonality between subcarriers is that they have constant amplitude, but differ only in phase. Hence, if any subcarrier of interest lies along the slope of the low pass filter, we will have ICI. Therefore, we need to allow a safety zone, as it were, around each frame by inserting zeros to subcarriers around the edges of the OFDM symbol. During this process, we must ensure that all the subcarriers of interest lie within the pass band of the low pass filter. Furthermore, the communication spectrum is crowded. This means that no extraneous signal should exist beyond the pass band of the filter (i.e., along the slope). The zeros ensure this. The data sequence is then given to the OFDM section. This section is self-explanatory. One point to be noted here is that each user in the DS-CDMA section will share the entire lot of subcarriers with other users. The discrimination between users will only be possible in the DS-CDMA section of the receiver after descrambling and will be based on the orthogonality of the pseudonoise sequences and the Walsh coding. The analog signal coming from the digital-to-analog converter (DAC) is then frequency hopped in the SFH section, before being fed to the RF transmitter. The hop set for each user usually bears a definite relationship with the pseudonoise sequence of a particular user. During this process, it must be ensured that:

- The frequency synthesizer of the hopper and the carrier beat frequency oscillator of the RF transmitter are phase locked;
- The frequency synthesizer of the dehopper and the beat frequency oscillator of the receiver (whose operating frequency is controlled by the synchronization circuit) are also phase locked.

Failure to ensure these two aspects will result in ICI. The hybrid receiver is exactly the reverse operation. The dehopped signal is given to the analog-to-digital converter (ADC), and thereafter, the digital signal processing starts with a training phase to determine the symbol timing and frequency offset. An FFT is used to demodulate all the subcarriers. The output of the OFDM section is the DS-CDMA sequence, which is then descrambled. The output of the DS-CDMA section is the QAM sequence, which are mapped onto binary values and decoded to produce binary output data. In order to successfully map the QAM values onto binary values, first the reference phases and amplitudes of all subcarriers have to be acquired. Alternatively, differential techniques can be applied.

3.5.2 Observations and Discussions

A detailed treatment of hybrid systems is given [6]. We make the following observations:

1. The MC-CDMA system, whether it is implemented as OFDM/ CDMA system or in any other manner, is extremely complex, as it involves spreading each bit across the available bandwidth. Our approach, on the other hand, spreads the entire word *before* the serial-to-parallel converter, thereby avoiding the complication of single bit spreading as is done in MC-CDMA systems. This means that we spread in the time domain, whereas MC-CDMA systems spread in the frequency domain to achieve the same final result. The MC-CDMA system, moreover, cannot cater to SFH to reduce near-far effect, because in such a case the system will become even more complex. In our approach, on the other hand, this can easily be done. Hence, we choose to call our approach the *Hybrid OFDM/CDMA/SFH* approach

or *Hybrid approach* for short. This approach has not been suggested before and it was first presented in the ACTS summit of 1999.

2. The advantage of the Hybrid approach lies in its easy implementation as compared to MC-CDMA, since the hardware complexity is less. This aspect has been examined in [6].

3. The problems of synchronization are transferred from the CDMA to the OFDM modulator. It is expected that it will be easier to solve the synchronization of the OFDM modulator as compared to a CDMA modulator, because the former uses cyclic prefixes, which can be exploited to achieve synchronization.

4. This proposal pertains to the downlink as well as to the uplink, the only difference being that for the synchronous downlink we can apply orthogonal Walsh-Hadamard sequences leading to the well-known user separation for MC-CDMA systems. On the other hand, in the asynchronous uplink scenario, pseudo-noise-sequences are used with the drawback of high MAI.

5. Table 3.2 summarizes the overall system aspects.

3.6 VSF-OFCDM

In this section, the variable spreading factor–orthogonal frequency and code division multiplexing scheme proposed by NTT DoCoMo is considered. VSF-OFCDM is designed specifically for the downlink, aiming at providing optimum link capacity for both isolated cell and multicell environments. In general, it can be seen as an adaptive MC-CDMA scheme, where the spreading factor (SF) is adapted on the basis of the cell environment and the propagation properties. In VSF-OFCDM, spreading is performed in both time and frequency domain across OFDM symbols and subcarriers, respectively. This is referred to as the time-frequency localized (TFL) spreading, or 2D-spreading, technique [36]. VSF-OFCDM is considered as a modulation scheme and not a multiple access scheme, because spreading codes are introduced mainly to achieve robustness against multicell interference rather than to provide multiple access.

As a result, it must be employed in conjunction with a multiple access scheme, such as TDMA, FDMA, CDMA, or a combination of these schemes.

The aim of this section is to highlight the main differences between VSF-OFCDM and the MC-CDMA discussed in Section 3.3, as well as to discuss the adaptation of spreading configuration to various link conditions. Therefore, the remainder of the section is structured as follows. First, the two-dimensional spreading procedure will be explained, leading to the

Table 3.2

Hybrid System Overall Aspects

Problem	Solution
1. OFDM system only supports one user.	We use OFDM/CDMA, which supports multiple users.
2. CDMA does not permit very high data rates, owing to frequency selective fading at high data rates.	OFDM counters this by S/P conversion, allowing flat fading at subcarrier level.
3. CDMA system suffers from near-far effect in the uplink.	This is solved in the hybrid system using SFH.
4. CDMA systems use DS-SFH to control the near-far effect. But DS-SFH mostly supports noncoherent modulation owing to hopping at bit level. However, coherent modulation is possible, but maintaining phase coherence between hops is difficult.	The OFDM-FH system hops on frame basis allowing coherent modulation.
5. CDMA systems cannot indefinitely support multiple users due to MAI and SI problems caused by too many users.	The hybrid system allows any number of users by increasing the number of hops. Hence, Number of Users = Number per CDMA System ´ Number of Hops. Bandwidth should, however, be available.
6. CDMA poses synchronization problems at very high chip rates.	OFDM systems have an easier synchronization problem due to using cyclic prefixes.

transceiver structure of VSF-OFCDM. Then, with a starting point in different cell environments, a discussion of the selection of spreading factor will follow. This part will be backed by some of the results obtained in [37, 38]. Finally, the main conclusions will be drawn.

3.6.1 Two-Dimensional Spreading Procedure

Figure 3.15 illustrates the spreading principles: (a) frequency domain spreading, (b) time domain spreading, and (c) time-frequency localized spreading, all with the same spreading factor SF = SFtime × SFfreq = 8.

The frequency domain spreading corresponds to the MC-CDMA scheme, where data is spread by a specific sequence and transmitted on SFfreq parallel subcarriers. The time domain spreading corresponds to the MC-DS-CDMA scheme, where data is spread by a specific sequence and transmitted on a fixed subcarrier over SFtime consecutive OFDM symbols. Ideally, when using orthogonal codes, the transmitted symbols can be perfectly recovered using

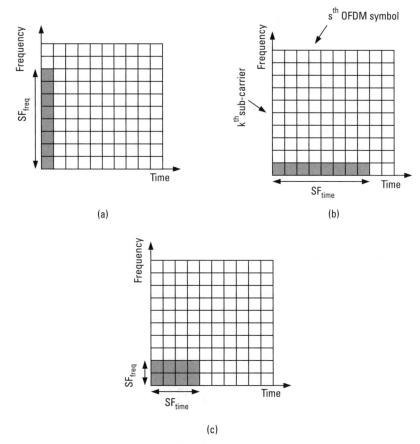

Figure 3.15 (a) Frequency domain spreading; (b) time domain spreading; and (c) time-frequency localized spreading. All with SF = 8.

preferably a simple receiver when considering the downlink scenario. However, in practice the code orthogonality is destroyed to some extent because the individual chips are scaled differently by the channel due to frequency selectivity and/or Doppler effect. This causes undesired intercede interference and finally leads to poor quality data estimates. Subsequently in this chapter the well-known term MAI will be used to denote intercode interference. The idea behind the time-frequency localized spreading (2D-spreading) is to utilize the fact that the channel may be locally approximated as flat fading (i.e., the channel is correlated in time and frequency) [35]. Therefore, localizing the spreading in time and frequency will cut down the difference in scaling of the individual chips and hence reduce MAI. In VSF-OFCDM, the two-dimensional spreading factor values are adaptively controlled according to the radio link conditions. The selection criteria are discussed in Section 3.6.3.

3.6.2   Transceiver Architecture

The transceiver structure employing VSF-OFCDM modulation with two-dimensional spreading is shown in Figure 3.16. At the transmitter (a), the incoming bits are first encoded, interleaved, and modulated into a sequence of data symbols. Then the data symbols are multiplied by a spreading code. The resulting chips are fed to a block shaper that assigns the chips to SFfreq selected subcarriers at SFtime successive OFDM symbols, depending on the setting of the variable SF controller. At the receiver the relevant chips are extracted from different OFDM symbols by the block deshaper. After despreading, the estimated data symbol sequence is demodulated, deinterleaved, and decoded. Analogous to MC-CDMA, different despreading operations such as EGC, MRC, or MMSEC may be utilized.

3.6.3   Spreading Configuration

In VSF-OFCDM the spreading factor controller of Figure 3.15 is adaptively configured. The configuration relies on several link condition parameters, namely, the intercell interference, the delay spread, and Doppler frequency, as

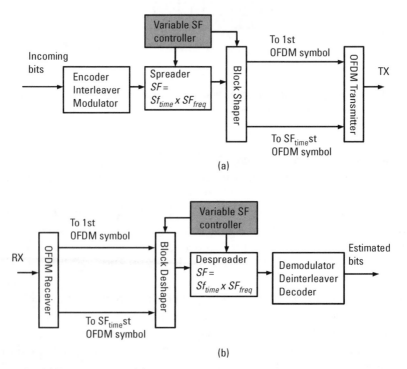

Figure 3.16 (a) Transmitter and (b) receiver.

well as on the major link parameters such as the data modulation scheme and coding rate [38].

3.6.3.1 Dependency on Intercell Interference

The key objective of VSF-OFCDM is to be able to operate efficiently in both isolated cell and multicell environments, which is fulfilled by adjusting the spreading factor configuration. In single-cell scenarios, such as hotspot areas or indoor offices, VSF-OFCDM employs SF = 1 (thus, OFDM) or a low SF value to avoid the potential MAI problem due to frequency selective fading or Doppler effect. On the other hand, in scenarios where intercell interference is present, a higher SF value is desirable as it introduces robustness against the intercell interference. The choice of SF value, in these cases, is a trade-off between reducing intercell interference and increasing MAI to achieve optimum link capacity [37].

By applying the spreading codes for multicell environments, VSF-OFCDM allows the one-cell frequency reuse scheme to be employed (see Figure 3.17). With the one-cell frequency reuse model, each cell uses the entire system bandwidth B and transmits using the same carrier frequency (f_c). In [37], a link capacity comparison for multicell environments was made between VSF-OFCDM employing one-cell frequency reuse and OFDM with three-cell frequency reuse.

For the three-cell frequency reuse model, each cell utilizes only one-third of the system bandwidth (i.e., $B/3$, and hence a total of three carrier frequencies are available for all cells). As a result, it is possible to create a cell structure where no adjacent cells use the same carrier frequency. This is the reason why intercell interference is weak in the three-cell reuse model and no spreading is required. The study shows that VSF-OFCDM with one-cell frequency reuse can achieve a link capacity gain of approximately 1.4 times compared to the OFDM with three-cell frequency reuse. In VSF-OFCDM the system load (i.e., the number of multiplexed codes in relation to the SF-value) is limited due to the increase of MAI. This explains the gain of only 1.4 times associated with a bandwidth gain of 3 times.

3.6.3.2 Dependency on the Link Parameters

The spreading factor also depends on the propagation characteristics—namely, the delay spread and Doppler spread of the channel—and on the other link parameters, such as modulation scheme and coding rate. When the wireless channel is more correlated in time than in frequency, it is suggested to prioritize time spreading [38]. This allows the time spreading to be applied without being penalized by MAI, especially for isolated cell environment, where the user mobility is low. Therefore, time spreading can be introduced in both isolated cell and multicell environments to facilitate specific features. For instance,

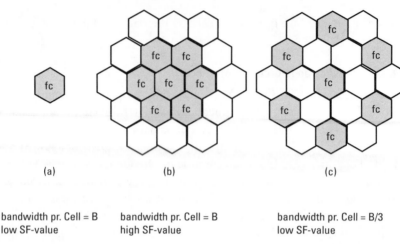

(a)	(b)	(c)
bandwidth pr. Cell = B	bandwidth pr. Cell = B	bandwidth pr. Cell = B/3
low SF-value	high SF-value	low SF-value

Figure 3.17 Cell environments: (a) isolated cell (hotspot or indoor office); (b) cellular with one-cell frequency reuse (multicell); and (c) cellular with three-cell frequency reuse (multicell).

by introducing time domain spreading within the same OFDM frame (i.e., without incurring any additional transmission delay), the control channel can be transmitted in parallel with the data channel using code multiplexing, which is an important requirement for adaptive modulation and coding (AMC) and hybrid automatic repeat request (HARQ) schemes. However, if the degree of orthogonality destruction in the time domain is significant due to high mobility, the spreading factor controller shall decide to spread less in the time domain [38].

At the cell border, when the intercell interference is strong, and when a low constellation modulation and a low channel coding rate is utilized, it is desirable to employ frequency domain spreading along with time domain spreading and interleaving. In this case the MAI increase due to the frequency domain spreading is insignificant as compared to the frequency diversity gain [38].

3.6.4 Summary

The VSF-OFCDM modulation scheme is originally based on MC-CDMA with an adaptive spreading factor extension. It has been further extended to support time domain spreading and is thus flexible in the spreading shape spanning from the general two-dimensional TFL spreading, through the one-dimensional frequency or time spreading (MC-CDMA or MC-DS-CDMA), to the special no-spreading case corresponding to normal OFDM. The spreading configuration is adapted on the basis of the propagation properties and the cell environment (i.e., in order to maximize the link capacity). In that sense the scheme offers deployment using the same air interface in isolated cell (hotspot or indoor office) and multicell (cellular systems) environments. For multiple access support, the scheme may be employed with several standard methods such as TDMA, FDMA (through subcarriers), CDMA, or combinations of these.

3.7 Conclusions

This chapter provided a summary of leading multiple access techniques and their classification into three types: contentionless (or scheduling), contention (or random), and hanging (or swimming). The hanging multiple access protocols include the multicarrier techniques. Multicarrier techniques (OFDM) successfully combat channel impairments and offer reasonably complex implementation to support high wireless data rates. Therefore multicarrier systems are the primary choice in many of the present 4G research efforts.

In this chapter, multiaccess schemes using OFDM have been considered: MC-CDMA, OFDMA, Hybrid OFDM-CDMA-SFH, and VSF-OFCDMA. The multiple access method employed in each scheme has been clearly extracted through time-frequency diagram. Furthermore, the specific features of each particular scheme have been highlighted. The chapter presented a state-of-the-art survey of the most important multi access schemes using OFDM. It represents a solid basis for identification of problems and viable solutions for the 4G development.

References

[1] Prasad R, *CDMA for Wireless Communications*, Norwood, MA: Artech House, 1996.

[2] Prasad R, *Universal Wireless Personal Communications*, Norwood, MA: Artech House, 1998.

[3] van Nee, R., and R. Prasad, *OFDM for Wireless Multimedia Communications*, Norwood, MA: Artech House, 2000.

[4] Ojanpera, T., and R. Prasad, (Eds.), *WCDMA: Towards IP Mobility and Mobile Internet*, Norwood, MA: Artech House, 2001.

[5] Hara, S., and R. Prasad, *Multicarrier Techniques for 4G Mobile Communications*, Norwood, MA: Artech House, 2003.

[6] Prasad, R., *OFDM for Wireless Communications Systems*, Norwood, MA: Artech House, 2000.

[7] ARIB FPLMTS Study Committee, "Report on FPLMTS Radio Transmission Technology SPECIAL GROUP (Round 2 Activity Report)," Draft v.E1.1, January 1997.

[8]   Chuang, J., and N. Sollenberger, "Beyond 3G: Wideband Wireless Data Access Based on OFDM and Dynamic Packet Assignment," *IEEE Communication Magazine*, Vol. 38, No. 7, July 2000, pp. 78–87.

[9]   Rom, R., and M. Sidi, *Multiple Access Protocols Performance and Analysis*, New York: Springer-Verlag, 1990.

[10]   Sunshine, C. A., *Computer Network Architecture and Protocols*, New York: Plenum Press, 1989.

[11]   Chouly, A., A. Brajal, and S. Jourdan, "Orthogonal Multicarrier Techniques Applied to Direct Sequence Spread Spectrum CDMA Systems," *Proc. of IEEE GLOBECOM'93*, November 1993, pp.1723–1728.

[12]   Yee, N., J.-P. Linnartz, and G. Fettweis, "Multi-Carrier CDMA in Indoor Wireless Radio Networks," *Proc. of IEEE PIMRC'93*, September 1993, pp.109–113.

[13]   Hara, S., et al., "Transmission Performance Analysis of Multi-Carrier Modulation in Frequency Selective Fast Rayleigh Fading Channel," *Wireless Personal Communications*, Vol. 2, No. 4, 1995/1996, pp. 335–356.

[14]   Monsen, P., "Digital Transmission Performance on Fading Dispersive Diversity Channels," *IEEE Trans. Commun.*, Vol. COM-21, January 1973, pp. 33–39.

[15]   Proakis, J. G., *Digital Communications*, Third Edition, New York: McGraw Hill, 1995, pp. 758–785.

[16]   Rappaport, T. S., *Wireless Communications, Principles and Practice*, Upper Saddle River, NJ: Prentice-Hall, 1996, pp. 188–189.

[17]   Steele, R., *Mobile Radio Communications*, London: Prentech Press, 1992, pp. 727–729.

[18]   Cimini, L. J., "Analysis and Simulation of a Digital Mobile Channel Using Orthogonal Frequency Division Multiplexing," *IEEE Trans. on Commun.*, Vol. COM-33, No. 6, June 1985, pp. 665–675.

[19]   Sampei, S., *Applications of Digital Wireless Technologies to Global Wireless Communications*, Upper Saddle River, NJ: Prentice-Hall, 1997, pp. 315–332.

[20]   Duel-Hallen, A., J. Holtzman, and Z. Zvonar, "Multiuser Detection for CDMA System," *IEEE Personal Communications*, Vol. 2, No. 2, 1995, pp. 46–58.

[21]   Rohling, H., et al., "Performance Comparison of Different Multiple Access Schemes for the Downlink of an OFDM Communication System," *VTC '97*, 1997, pp. 1365–1369.

[22]   Chuang, J. C.-I., "An OFDM-Based System with Dynamic Packet Assignment and Interference Suppression for Advanced Cellular Internet Service," *Globecom'98*, Sydney, Australia, 1998, Vol. 2, pp. 974–979.

[23]   Pottie, G. J., and A. R. Calderbank, "Channel Coding Strategies for Cellular Radio," *IEEE International Symposium on Information Theory*, San Antonio, TX, January 1993, pp. 251.

[24]   Viterbi, A. J., "The Orthogonal-Random Waveform Dichotomy for Digital Mobile Communication," *IEEE Personal Communications*, First Quarter 1994, pp.18–24.

[25]   van Nee, R., and R. Prasad, *OFDM for Wireless Multimedia Communications*, Norwood, MA: Artech House, 2000.

[26] Suzuki, M., R. Boehnke, and K. Sakoda, "BDMA—Band Division Multiple Access: New Air-Interface for 3rd Generation Mobile System, UMTS, in Europe," *Proceedings ACTS Mobile Communications Summit*, Aalborg, Denmark, October 1997, pp. 482–488.

[27] Olofsson, H., J. Naslund, and J. Sköld, "Interference Diversity Gain in Frequency Hopping GSM," *Proceedings IEEE Vehicular Technology Conference*, Chicago, June 1995, pp. 102–106.

[28] "ETSI Draft Specification on the Selection Procedure for the Choice of Radio Transmission Technologies of the Universal Mobile Telecommunication System (UMTS)," ETR/SMG-50402, Ver. 0.9.5.

[29] Jankiraman, M., and R. Prasad, "Hybrid CDMA/OFDM/SFH: A Novel Solution for Wideband Multimedia Communications," ACTS Summit, Sorento, Italy, 1999.

[30] Jankiraman, M., and R, Prasad, "A Novel Algorithmic Synchronization Technique for OFDM Based Wireless Multimedia Communications," ICC '99, 1999, Vol. 1, pp. 528–533, Vancouver, Canada..

[31] Jankiraman, M., and R. Prasad, "Hybrid CDMA/OFDM/SFH: A Novel Solution for Wideband Multimedia Communications," MC-SS '99, Oberpffaffenhausen, Germany, 1999.

[32] Jankiraman, M., and R. Prasad, "A Novel Solution to Wireless Multimedia Application: The Hybrid OFDM/CDMA/SFH Approach," PIMRC 2000, 2000, Vol.2, pp. 1368–1374, London.

[33] Jankiraman, M., and R. Prasad, "Algorithm Assisted Synchronization of OFDM Systems for Fading Channels," 5th International OFDM-Workshop, Hamburg, Germany, September 2000.

[34] Jankiraman, M., and R. Prasad, "Performance Evaluation of Hybrid OFDM/CDMA/SFH for Wireless Multimedia," VTC 2000, 2000, Vol. 2, pp. 934–941 Boston, MA.

[35] Jankiraman, M., and R. Prasad, "Wireless Multimedia Application: The Hybrid OFDM/CDMA/SFH Approach," ISSSTA 2000, 2000, Vol. 2, pp. 387–393 Piscataway, NJ.

[36] Persson, A., T. Ottosson, and E. Ström, "Time-Frequency Localized CDMA for Downlink Multi-Carrier Systems," in IEEE ISSSTA, Vol. 1, Prague, September 2002, pp. 118–122.

[37] Atarashi, H., S. Abeta, and M. Sawahashi, "Variable Spreading Factor-Orthogonal Frequency and Code Division Multiplexing (VSF-OFCDM) for Broadband Packet Wireless Access," IEICE Trans. Commun., Vol. E86-B, No. 1, January 2003, pp. 291–299.

[38] Kishiyama, Y., et al., "Experiments on Throughput Performance Above 100-Mbps in Forward Link for VSF-OFCDM Broadband Wireless Access," in IEEE VTC, Vol. 3, Orlando, Florida, October 2003, pp. 1863–1868.

4

Emerging Technologies for 4G

4.1 Introduction

OFDM-based access techniques are the most appropriate candidate for 4G, as it has been discussed in Chapter 3. In order to provide the required quality of services to users of 4G, several key technologies will be used with OFDM. Therefore, the next task is to present and update information on key technologies, namely MIMO technology, radio resource management, software defined radio (SDR) communication system, mobile IP, and relaying techniques [1-175]. This chapter starts with the multiantenna technologies in Section 4.2. Multiantenna is one of the key technologies for mitigating the negative effects of the wireless channel, providing better link quality and/or higher data rate without consuming extra bandwidth or transmitting power. The use of multiple antennas at either receiver, transmitter, or both ends provides several benefits: array gain, interference reduction, diversity gain, and/or multiplexing gain. Section 4.3 focuses on the main radio resource management functions. They constitute a fundamental aspect for the provision of a variability degree of the quality of service and the effective explanation of the limited radio resource, thus representing a hot research area. The concept of SDR is one of the bridges between software programming and real hardware implementation. Section 4.4 defines radio communication systems and summarizes the advantages of SDR. In addition, this section discusses the problems that must be overcome to realize the SDR communication system and explains the remarkable technologies developed to realize such a system. Section 4.5 deals with IP network issues: Mobile IP architecture, proposed for supporting mobility in Internet, is presented together with guidelines behind mobility. Section 4.6 introduces the concepts of relaying techniques for 4G, and other enabling techniques are listed in Section 4.7.

4.2   Multiantenna Technologies

Recent years have witnessed an explosive growth of interest in studying and using multiple antenna techniques for wireless communication systems. Such a trend is motivated by the fact that several of the specifications foreseen for future wireless systems appear to be difficult, if not impossible, to fulfill with conventional single antenna systems. It is widely accepted that multiple antennas have the potential to increase the achievable data throughput, to enhance link quality (BER, QoS), and to increase cell coverage and network capacity, among others. Such a promising array of enhancements has contributed to speeding up the development in the field, both at academic levels, where countless techniques are developed, and in the industry, where solutions based on these techniques are being rapidly adopted in real systems.

Multiple transmit and receive antennas can be combined with various multiple access techniques such as TDMA, CDMA, and OFDM to improve the capacity and reliability of communications. Multiple-input multiple-output (MIMO) communication systems are regarded as an effective solution for future high-performance wireless networks. The use of multiple antennas at transmitter and receiver, popularly known as MIMO, is a promising cost-effective technology that offers substantial leverages in making the anticipated 1-Gbps wireless links a reality.

Several efforts are currently underway to build non-line-of-sight (NLOS) broadband wireless systems. A MIMO wireless system (physical layer and MAC layer technology) using OFDM modulation for NLOS environments was successfully developed by NTT DoCoMo. In mobile access, there is an effort under the ITU working group to integrate MIMO techniques into the high-speed downlink packet access (HSDPA) channel, which is part of the Universal Mobile Telecommunications System (UMTS) standard. Lucent Technologies recently announced a chip for MIMO enhancement of UMTS/HSDPA but has released no further details. Preliminary efforts are also underway to define a MIMO overlay for the IEEE 802.11 standard for WLANs under the newly formed Wireless Next Generation (WNG) group. Moreover, many other companies also proposed advanced MIMO schemes for the IEEE 802.16 standard. In this chapter, we provide an overview of MIMO technology and explain recent research directions and results.

4.2.1   Overview of MIMO Technology

Depending on the geometry of the employed antenna array, two basic multiantenna approaches can be considered: a beamforming approach for closely separated antenna elements (interelement separation is at most $\lambda/2$, where λ is the carrier wavelength) or a diversity approach for widely separated antenna elements (typical interelement spacing is at least a few λ). In this chapter, we will explore the latter approach where the fading processes associated with any two possible transmit-receive antenna pair can be assumed to be independent. The fact that a MIMO system consists of a number of uncorrelated concurrent channels has been exploited from two different perspectives. First, from a pure diversity standpoint, one can enhance the fading statistics of the received signal by virtue of the multiple available replicas being affected by independent fading channels. By sending the same signal through parallel and independent fading channels, the effects of multipath fading can be greatly reduced, decreasing the outage probability and hence improving the reliability of the communication link [1]. In the second approach, referred to as spatial multiplexing [2], different information streams are transmitted on parallel spatial channels associated with the transmit antennas. This could be seen as a very effective method to increase spectral efficiency. In order to be able to separate the individual streams, the receiver has to be equipped with at least as many receive antennas as the number of parallel channels generated by the transmitter in general. For a given multiple antenna configuration, one may be interested in finding out which approach would provide the best possible or desired performance.

4.2.1.1 Diversity

Space-time coding (STC) is a hybrid technique that uses both space and temporal diversity in a combined manner. There are two forms of STC namely space-time block code (STBC) and space-time trellis code (STTC). STBC efficiently exploits transmit diversity to combat multipath fading while keeping decoding complexity to a minimum. Tarokh et al. [3] showed that no STBC can achieve full-rate and full-diversity for more than two transmit antennas, and proposed a 3/4 rate, full-diversity code for four transmit antennas. A full-rate quasi-orthogonal (QO) STBC was proposed by Jafarkhani [4] for four transmit antennas based on Alamouti orthogonal STBC [1]. In this case, the transmission matrix is given by

$$
\mathbf{C} = \begin{bmatrix} \mathbf{A}_{12} & \mathbf{A}_{34} \\ -\mathbf{A}_{34}^* & \mathbf{A}_{12}^* \end{bmatrix} = \begin{bmatrix} x_1 & x_2 & x_3 & x_4 \\ -x_2^* & x_1^* & -x_4^* & x_3^* \\ -x_3^* & -x_4^* & x_1^* & x_2^* \\ x_4 & -x_3^* & -x_2^* & x_1 \end{bmatrix} \tag{4.1}
$$

where $\mathbf{A}_{12}$, $\mathbf{A}_{34}$ are the Alamouti codes [1]. It is noted here that since the channel matrix of the QO-STBC is not full-rank, full-diversity gain cannot be attained. To achieve the full-diversity and full-rate (FDFR) property, a new FDFR STC approach was recently proposed.

4.2.1.2 Mixed (Hybrid Diversity and Spatial Multiplexing)

This mode combines diversity and spatial multiplexing by transmitting from four transmit antennas, each space-time block coded with the basic Alamouti scheme of order two [1]. The transmission matrix of the space-time block coding for the ith data stream, $i = a,b$ is given by

$$\mathbf{A}_i = \begin{bmatrix} x_1(i) & x_2(i) \\ -x_2(i)^* & x_1(i)^* \end{bmatrix} \tag{4.2}$$

To decode the data, minimum mean square error (MMSE) and zero forcing (ZF) receivers can be employed. For the MMSE receiver, we assume that the transmitted matrix is $\left[a_{n2}(k), a_{2n+1}(k), b_{2n}(k), b_{2n+1}(k)\right]^T$, where a and b indicate different signal streams. First, the tap weight vector and decoding layer order are determined. If the first decoding layer is a, the procedure can be represented by

$$\begin{bmatrix} \hat{a}_{2n}(k) \\ \hat{a}_{2n+1}(k) \end{bmatrix} = decision\left\{ \begin{bmatrix} \mathbf{w}_1^H(k) \\ \mathbf{w}_2^H(k) \end{bmatrix} \mathbf{y}_n(k) \right\} \tag{4.3}$$

The interference from the original signal can be subtracted using $\hat{a}_{2n}(k)$ and $\hat{a}_{2n+1}(k)$; accordingly, the other stream can be decoded as follows:

$$\mathbf{y}_n'(k) = \mathbf{y}_n(k) - \left[\mathbf{h}_1(k) \quad \mathbf{h}_2(k)\right] \begin{bmatrix} \hat{a}_{2n}(k) \\ \hat{a}_{2n+1}(k) \end{bmatrix}$$

$$\begin{bmatrix} \hat{b}_{2n}(k) \\ \hat{b}_{2n+1}(k) \end{bmatrix} = decision\left\{ \begin{bmatrix} \mathbf{h}_3(k) \\ \mathbf{h}_4(k) \end{bmatrix} \mathbf{y}_n'(k) \right\} \tag{4.4}$$

Note that for comparative purposes we can also employ maximum likelihood (ML) decoding (explained in next section) to obtain the optimum performance, which was used as our baseline reference.

4.2.1.3 Spatial Multiplexing

In this section, we briefly review the spatial multiplexing scheme. The vertical Bell Lab layered space time (V-BLAST) architecture has been recently proposed for achieving high spectral efficiency over wireless channels characterized by rich scattering [2]. In this approach, one way of detection is to use conventional adaptive antenna array (AAA) techniques (i.e., linear combining and nulling). Conceptually, each stream (i.e., layer) in turn is considered to be the desired signal, while regarding the remaining signals as interference. Nulling is performed by linearly weighting the received signals so as to satisfy some performance related criterion, such as ZF or MMSE. This linear nulling approach is viable, but superior performance is obtained if nonlinear techniques are used. One particularly attractive nonlinear alternative is to exploit symbol cancellation as well as linear nulling to perform detection. By using symbol cancellation, the interference from the already-detected components is subtracted from the received signal vector, effectively reducing the overall interference. Here we will consider ordered successive interference cancellation with ZF and MMSE. Also, a ML decoding receiver will be used as a reference.

It is assumed that the $H_{ij}(k)$ is the channel coefficient from j_{th} transmit antenna to i_{th} receive antenna and $\mathbf{w}$ is white Gaussian noise with covariance matrix $\mathbf{C}_w = E\left[\mathbf{ww}^H\right] = \sigma^2\mathbf{I}_{NR}$ where N_R is the number of received antennas. Then, the received signal vector can be written as follows:

$$\mathbf{y}_n(k) = \mathbf{H}(k)x_n(k) + \mathbf{w}(k) \tag{4.5}$$

where the index k denotes the kth subcarrier, $\mathbf{y}(k) = \left[y_1(k)\cdots y_{N_R}(k)\right]^T$, $\mathbf{x}(k) = \left[x_1(k)\cdots x_{N_R}(k)\right]^T$, and $\mathbf{w}(k)$ is the $(N_R \times 1)$ noise vector.

Maximum Likelihood Decoding (Optimal Solution)

The ML detection of $\mathbf{x}(k)$ can be found by maximizing the conditional probability density function and this is equivalent to minimizing the log-likelihood function:

$$\hat{\mathbf{x}}(k) = \min_{\mathbf{x}(k)}\left\{\mathbf{y}(k) - \mathbf{Hx}(k)\right\}^H\left\{\mathbf{y}(k) - \mathbf{Hx}(k)\right\} \tag{4.6}$$

where, $\mathbf{x}(k) \in$ all possible constellation sets.

It is well known that ML decoding is characterized by a high implementation complexity, and thus, suboptimal but practically implementable solutions are considered next.

Ordered Successive Interference Cancellation (OSIC)

Instead of the ML decoding approach, linear detection techniques can be used (i.e., ZF and MMSE). To improve the linear detection techniques, we try to decode according to received signal strength, and extract the decoded signal from the received signal. This approach is referred to as D-BLAST or V-BLAST [2] according to the transmitted signal structure, where D stands for diagonal and V for vertical. For simplicity, we consider the OSIC.

The receiving operation of OSIC can be summarized as follows:

- Step 1: Compute the tap weight matrix W.
- Step 2: Find the layer with maximum SNR.
- Step 3: Detection

$$z_k(n) = \mathbf{W}_k^H \mathbf{y}(n)$$

$$\hat{x}_k(n) = decision[z_k(n)].$$

- Step 4: Interference cancellation

$$\mathbf{y}(n) = \mathbf{y}(n) - \hat{\mathbf{h}}_k(n)$$

$$\mathbf{H} = [\mathbf{h}_1, \cdots, \mathbf{h}_{k-1}, 0, \mathbf{h}_{k+1}, \cdots, \mathbf{h}_T]$$

- Step 5: Repeat Step 1 to 5 until all symbols are detected.

Zero-Forcing.   The cost function can be expressed as

$$J_{ZF} = \{\mathbf{y}(k) - \mathbf{H}\hat{x}(k)\}^H \{\mathbf{y}(k) - \mathbf{H}\hat{x}(k)\} \tag{4.7}$$

Since J_{ZF} is a convex function over $\hat{x}(k)$, $\hat{x}(k)$ can be determined by using the minimum limit. Then, the tap weight vector is given by

$$\mathbf{W} = \{\mathbf{H}^H \mathbf{H}\}^{-1} \mathbf{H}^H \tag{4.8}$$

Minimum Mean Square Error. To take into account the noise variance, the cost function can be expressed as

$$J_{MMSE} = E\left[\{\mathbf{y}(k) - \mathbf{H}\hat{x}(k)\}^H \{\mathbf{y}(k) - \mathbf{H}\hat{x}(k)\}\right] \tag{4.9}$$

Using a similar method as in the ZF detection method, the weight vector results in

$$\mathbf{W} = \{\mathbf{H}^H + \sigma^2 I\}^{-1} \mathbf{H}^H \tag{4.10}$$

Note that the noise variance σ^2 has to be estimated in order to use the MMSE approach.

4.2.2 Adaptive Multiple Antenna Techniques

Recently some authors have considered the diversity-spatial multiplexing problem. In [5], the fundamental trade-off between diversity and spatial multiplexing is explored by Zheng and Tse. A scheme based on switching between diversity and spatial multiplexing is proposed by Heath and Paulraj [6]. Authors have considered a fixed rate system in which the receiver adaptively selects one of the two transmission approaches based on the largest minimum Euclidean distance of the received constellation. The receiver informs its selection to the transmitter via a 1-bit feedback channel. To ensure a fixed bit rate, the diversity scheme uses modulation with a higher order than that used by its counterpart spatial modulation case. Skjevling et al. presented a hybrid method combining both diversity and spatial multiplexing [7]. The proposed approach optimally assigns antennas to a given (fixed) transmission scheme combining diversity and spatial multiplexing. Antenna selection is based either on full channel feedback or long-term statistics. Gorokhov et al. studied the relationship between multiplexing gain and diversity gain in the context of antenna subset selection [8], thereby extending the recent result by Zheng and Tse [5].

4.2.3 Open-Loop MIMO Solutions

Alamouti developed a remarkable orthogonal FDFR code for $N_T = 2$ transmit antennas [1], requiring a simple linear decoder at the receiver. Tarokh et al. [3] proved that a FDFR orthogonal code only exists for $N_T = 2$ and proposed some space-time block codes for $N_T > 2$ attaining full-diversity but not full-rate. In [4] a quasi-orthogonal full-rate code is proposed by Jafarkhani, though full-diversity gain cannot be attained. Based on space-time constellation rotation, Xin et al. [9] and Ma et al. [10] proposed a FDFR encoder for an arbitrary number of transmit antennas. For an even number of transmit antennas, Jung et al. [11] obtained coding gain with a FDFR space-time block code by serially concatenating the Alamouti scheme with the constellation rotation techniques used in [9, 10]. Although the Alamouti-based space-time constellation rotation encoder (A-ST-CR) of [11] can effectively achieve full-diversity and full-rate, the decoding complexity is an issue and its practical implementation becomes prohibitive, even for a small number of transmit antennas (e.g., $N_T = 4$). This is in virtue of the high computational complexity required by the ML decoding algorithm.

In addressing the complexity problem, this chapter further extends the results of [11] by considering a system based on the serial concatenation of a new rotating precoding scheme with the basic Alamouti codes of order two. A proper process of puncturing and shifting after the actual constellation-rotation operation can conveniently decompose the encoding process into rotation operations carried out in a lower order matrix space. The impact of this puncture and shift

rotation coding scheme is very significant at the receiver, where, due to the provided signal decoupling, the decoding complexity is significantly reduced. It is
shown in this chapter that the proposed method attains the same performance as
the scheme presented in [11] with a substantial complexity reduction.

References [9-11] use a precoder based on the Vandermonde matrix for
attaining a FDFR system. After multiplying the received signal x by the
Vandermonde matrix, each component of vector r combines all the symbols as
can be observed in the next basic precoder equation.

$$\mathbf{r} = \Theta\mathbf{x} = \frac{1}{\sqrt{4}} \begin{bmatrix} 1 & \alpha_0^1 & \alpha_0^2 & \alpha_0^3 \\ 1 & \alpha_1^1 & \alpha_1^2 & \alpha_1^3 \\ 1 & \alpha_2^1 & \alpha_2^2 & \alpha_2^3 \\ 1 & \alpha_3^1 & \alpha_3^2 & \alpha_3^3 \end{bmatrix} \begin{bmatrix} x_1 \\ x_2 \\ x_3 \\ x_4 \end{bmatrix} = \begin{bmatrix} r_1 \\ r_2 \\ r_3 \\ r_4 \end{bmatrix} \tag{4.11}$$

where $\alpha_i = \exp\left(j2\pi(1+1/4)/N\right)$, $i = 0, 1, \ldots, N-1$.

Xin [9] and Ma [10] use a diagonal channel matrix after multiplying the
information symbols by the Vandermonde matrix. This linear precoding is
referred to as the constellation rotation operation. Notice that the coding advantage of [9, 10] is not optimized, although the schemes successfully achieve
FDFR. Jung [11] improves the coding advantages by concatenating the constellation rotating precoder with the basic Alamouti scheme, resulting in the
following transmitted signals:

$$\mathbf{S} = \begin{bmatrix} r_1 & r_2 & 0 & 0 \\ -r_2^* & r_1^* & 0 & 0 \\ 0 & 0 & r_3 & r_4 \\ 0 & 0 & -r_4^* & r_3^* \end{bmatrix} \tag{4.12}$$

At the receiving end, the signal can be written as

$$\mathbf{y} = \begin{bmatrix} y_1 \\ y_2^* \\ y_3 \\ y_4^* \end{bmatrix} = \frac{1}{\sqrt{2}} \begin{bmatrix} h_1 & h_2 & 0 & 0 \\ h_2^* & -h_1^* & 0 & 0 \\ 0 & 0 & h_3 & h_4 \\ 0 & 0 & h_3^* & -h_4^* \end{bmatrix} \begin{bmatrix} r_1 \\ r_2 \\ r_3 \\ r_4 \end{bmatrix} + \begin{bmatrix} n_1 \\ n_2 \\ n_3 \\ n_4^* \end{bmatrix} = \mathbf{Hr} + \mathbf{n} \tag{4.13}$$

Note that since r_1, r_2, r_3, r_4 already sums over $x_1 \sim x_4$ through the
Vandermonde matrix, each symbol experiences the channel twice. We can now
point out that (4.12) can be separated into two parts: (r_1, r_3 and r_2, r_4) or (r_1, r_4

and r_2, r_3). Consequently, the Vandermonde matrix for the precoder need not be of size 4, but smaller. Based on this observation, we can use a puncturing and shifting operation after the constellation rotation process resulting in a new precoder.

$$
\mathbf{r}_{1,3} = \begin{bmatrix} r_1 \\ r_3 \end{bmatrix} = \frac{1}{\sqrt{2}} \begin{bmatrix} 1 & \alpha_0^1 \\ 1 & \alpha_2^1 \end{bmatrix} \begin{bmatrix} x_1 \\ x_3 \end{bmatrix}
\quad\text{or}\quad
\mathbf{r}_{1,4} = \begin{bmatrix} r_1 \\ r_4 \end{bmatrix} = \frac{1}{\sqrt{2}} \begin{bmatrix} 1 & \alpha_0^1 \\ 1 & \alpha_3^1 \end{bmatrix} \begin{bmatrix} x_1 \\ x_4 \end{bmatrix}
$$

$$
\mathbf{r}_{2,4} = \begin{bmatrix} r_2 \\ r_4 \end{bmatrix} = \frac{1}{\sqrt{2}} \begin{bmatrix} 1 & \alpha_1^1 \\ 1 & \alpha_3^1 \end{bmatrix} \begin{bmatrix} x_2 \\ x_4 \end{bmatrix}
\qquad
\mathbf{r}_{2,3} = \begin{bmatrix} r_2 \\ r_3 \end{bmatrix} = \frac{1}{\sqrt{2}} \begin{bmatrix} 1 & \alpha_1^1 \\ 1 & \alpha_2^1 \end{bmatrix} \begin{bmatrix} x_2 \\ x_3 \end{bmatrix}
\tag{4.14}
$$

After puncturing and shifting, the encoder can be defined as

$$
\frac{1}{\sqrt{2}} \begin{bmatrix} 1 & \alpha_0^1 & 0 & 0 \\ 0 & 0 & 1 & \alpha_1^1 \\ 1 & \alpha_2^1 & 0 & 0 \\ 0 & 0 & 1 & \alpha_3^1 \end{bmatrix}
\quad\text{or}\quad
\frac{1}{\sqrt{2}} \begin{bmatrix} 1 & \alpha_0^1 & 0 & 0 \\ 0 & 0 & 1 & \alpha_1^1 \\ 0 & 0 & 1 & \alpha_2^1 \\ 1 & \alpha_3^1 & 0 & 0 \end{bmatrix}
\tag{4.15}
$$

Recently, Zafar *et al.* proposed a low decoding complexity (symbol by symbol decoding) improved space-time code with full-diversity for three and four transmit antennas configurations [13]. The following is the format obtained after modifying the transmission matrix:

$$
\mathbf{s} = \begin{bmatrix} x_1 + jy_3 & -x_2 + jy_4 & 0 & 0 \\ x_2 + jy_4 & x_1 - jy_3 & 0 & 0 \\ 0 & 0 & x_3 + jy_1 & -x_4 + jy_2 \\ 0 & 0 & x_4 + jy_2 & x_3 - jy_1 \end{bmatrix}
\tag{4.16}
$$

where $x_i = s_{iI} \cos\theta - s_{iQ} \sin\theta$, $y_i = s_{iI} \cos\theta - s_{iQ} \sin\theta$, and $\theta = \tan^{-1}\left(\frac{1}{3}\right)$. The complex symbols s_i take values from a QAM signal set. Note that we already separated the encoder into two parts, so a Vandermonde matrix of order two should be used. The decoding computational complexity is significantly reduced compared to that of Jung's in [11].

4.2.4 Closed-Loop MIMO Solutions

In this section we explain closed-loop MIMO solutions, which consist of two parts: antenna grouping and codebook based schemes, which use feedback information from a mobile station.

4.2.4.1 Antenna Grouping

The rate 1 transmission code for four transmit-antenna base stations in the IEEE 802.16e is

$$A = \begin{bmatrix} s_1 & -s_2^* & 0 & 0 \\ s_2 & s_1^* & 0 & 0 \\ 0 & 0 & s_3 & -s_4^* \\ 0 & 0 & s_4 & s_3^* \end{bmatrix} \tag{4.17}$$

Note that this scheme does not achieve full-diversity.

The effective channel model H_{eff} is orthogonal and it can be written as follows:

$$H_{\text{eff}}^H H_{\text{eff}} = \begin{bmatrix} \rho_1 & 0 & 0 & 0 \\ 0 & \rho_1 & 0 & 0 \\ 0 & 0 & \rho_2 & 0 \\ 0 & 0 & 0 & \rho_2 \end{bmatrix}, \tag{4.18}$$

where $\rho_1 = |h_1|^2 + |h_2|^2$ and $\rho_2 = |h_3|^2 + |h_4|^2$, being $h_k, k = 1,2,3,4$ the channel coefficient associated with the kth transmit antenna.

If the BS can use channel state information, the performance of the existing matrix A approaches the performance of the FDFR STC in section 4.2.3 by using the following equation:

$$\arg \min_{antenna\_pair} |\rho_1 - \rho_2| \tag{4.19}$$

Let $d_{\min}^2$ be the corresponding minimum distance of the normalized unit energy constellation. The $2^R - \text{QAM}$ Euclidean distance equation $d_{\min}^2 = 12/(2^R - 1)$ will be used, corresponding to QAM modulation for diversity. Using this Euclidean distance equation, we can estimate the error probability as

$$P_e \leq N_e Q\left(\sqrt{\frac{E_s}{N_0}} d_{\min}^2\right) \tag{4.20}$$

where $d_{\min}^2$ is the squared Euclidean distance of the received signal, and N_e is the number of nearest neighbors in the constellation, which can be found for each proposed mapping scheme based on the channel coefficient matrix H,

$Q(x) = \frac{1}{2} erfc(x/\sqrt{2})$, where erfc is the complementary error function. For STC, the minimum distance of the diversity constellation at the receiver can be shown to be

$$d_{\min}^2(\mathbf{H}) \leq \frac{\min\left(\|\mathbf{H}\|_F^2(a,b), \|\mathbf{H}\|_F^2(c,d)\right)}{N_T} d_{\min}^2 \qquad (4.21)$$

where (a, b) and (c, d) are antenna grouping index and $\|\mathbf{H}\|_F$ is the Frobenius norm of matrix H. The details for derivation follow the derivation procedure of the maximum SNR criterion for code design.

Figure 4.1 shows the system block diagram, which makes use of a grouper to select the antenna pair based on feedback channel information from the MS.

The performance of the proposed scheme is shown in Figure 4.2. At BER = 10^{-3} point, the proposed scheme outperforms the conventional STC without antenna grouping by 3.5 dB.

The rate 2 transmission code for four transmit antennas in the current IEEE 802.16 standard [1] is

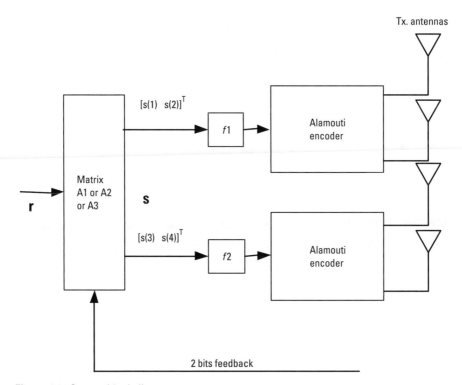

Figure 4.1 System block diagram.

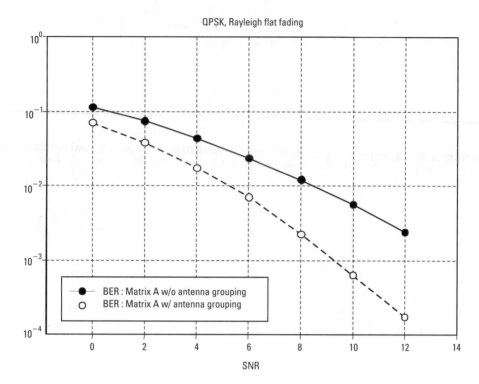

Figure 4.2 BER versus SNR with and without antenna grouping.

$$B = \begin{bmatrix} s_1 & -s_2^* & s_5 & -s_7^* \\ s_2 & s_1^* & s_6 & -s_8^* \\ s_3 & -s_4^* & s_7 & s_5^* \\ s_4 & s_3^* & s_8 & s_6^* \end{bmatrix} \tag{4.22}$$

Transmission matrix B for rate 2 can be improved with antenna grouping information. The BS can group antennas 0 and 1 for the first diversity pair and antennas 2 and 3 for the second diversity pair. In matrix form, this can be expressed as follows:

$$B_1 = \begin{bmatrix} s_1 & -s_2^* & s_5 & -s_7^* \\ s_2 & s_1^* & s_7 & s_5^* \\ s_3 & -s_4^* & s_6 & -s_8^* \\ s_4 & s_3^* & s_8 & s_6^* \end{bmatrix} \tag{4.23}$$

Considering different grouping index, the transmission matrix B can also be expressed as

$$B_2 = \begin{bmatrix} s_1 & -s_2^* & s_5 & -s_7^* \\ s_2 & s_1^* & s_7 & s_5^* \\ s_4 & s_3^* & s_8 & s_6^* \\ s_3 & -s_4^* & s_6 & -s_8^* \end{bmatrix} B_3 = \begin{bmatrix} s_1 & -s_2^* & s_5 & -s_7^* \\ s_3 & -s_4^* & s_6 & -s_8^* \\ s_2 & s_1^* & s_7 & s_5^* \\ s_4 & s_3^* & s_8 & s_6^* \end{bmatrix}$$

$$B_4 = \begin{bmatrix} s_1 & -s_2^* & s_5 & -s_7^* \\ s_4 & s_3^* & s_8 & s_6^* \\ s_2 & s_1^* & s_7 & s_5^* \\ s_3 & -s_4^* & s_6 & -s_8^* \end{bmatrix} B_5 = \begin{bmatrix} s_1 & -s_2^* & s_5 & -s_7^* \\ s_3 & -s_4^* & s_6 & -s_8^* \\ s_4 & s_3^* & s_8 & s_6^* \\ s_2 & s_1^* & s_7 & s_5^* \end{bmatrix} \qquad (4.24)$$

$$B_6 = \begin{bmatrix} s_1 & -s_2^* & s_5 & -s_7^* \\ s_4 & s_3^* & s_8 & s_6^* \\ s_3 & -s_4^* & s_6 & -s_8^* \\ s_2 & s_1^* & s_6 & s_5^* \end{bmatrix}$$

At the mobile, the optimum transmission matrix is determined based on the following criteria. Let Y_{ri} be the received signal at the ith symbol time at the rth receive antenna, and $h_{t,r}$ denote the channel parameter between the $h_{t,th}$ transmit and rth receive antenna. When the number of receive antennas is two, the received signal can be represented as

$$y = X(HW)s + v \qquad (4.25)$$

where $y = \begin{bmatrix} y_{1,1} & y_{1,2}^* & y_{2,1} & y_{2,2}^* \end{bmatrix}^T$, $s = \begin{bmatrix} s_1 & s_2 & s_3 & s_4 \end{bmatrix}^T$, v is the noise

vector, $H = \begin{bmatrix} h_{1,1} & h_{2,1} & h_{3,1} & h_{4,1} \\ h_{1,2} & h_{2,2} & h_{3,2} & h_{4,2} \end{bmatrix}$, $X(\cdot)$ is a function of 2-by-4 input matrix, which is defined as

$$X\left(\begin{bmatrix} a & b & c & d \\ e & f & g & h \end{bmatrix} \right) = \begin{bmatrix} a & b & c & d \\ b^* & -a^* & d^* & -c^* \\ e & f & g & h \\ f^* & -e^* & h^* & -g^* \end{bmatrix} \qquad (4.26)$$

and W is a permutation of matrix B.

At the mobile station, the index of the transmission matrix B_q, $q = 1,2,3,\ldots6$ is determined based on the following criteria:

$$q = \arg \min_{l=1,\ldots,6}\left[abs\left(\det(\mathbf{H}_{l,1}) + \det(\mathbf{H}_{l,2})\right)\right] \qquad (4.27)$$

where $\mathbf{H}_{l,1}$ is the first two columns of $\mathbf{HW}_l$, and $\mathbf{H}_{l,2}$ is the last two columns of $\mathbf{HW}_l$. Note that the antenna grouping matrix selection rule in (4.27) is equivalent to the following rule:

$$q = \arg \min_{l=1,\ldots,6}\left[trace\left(\left[\left(\mathbf{X}(\mathbf{HW}_l)\right)^H \mathbf{X}(\mathbf{HW}_l)\right]^{-1}\right)\right] \qquad (4.28)$$

Alternate criteria for the antenna grouping can be applied to determine antenna group index. For example, minimize BER, MMSE, and so forth.

We compare the proposed antenna grouping-based closed-loop STC with open-loop STC for four transmit-antenna rate 2 STC. In Figure 4.3, packet

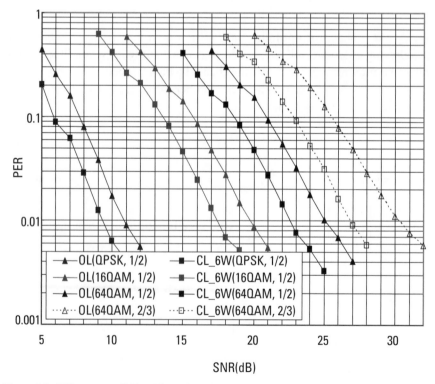

Figure 4.3 PER versus SNR with and without antenna grouping when the correlation coefficient is 0.7.

error rates (PERs) of the proposed antenna grouping method and the conventional open-loop STC (matrix B) method are compared in the pedestrian A channel with 3 km/h. One frame feedback delay is reflected in the simulation, and an MMSE linear detector is used at the receiver.

When the correlation coefficient is 0.7 (Figure 4.3), the proposed antenna grouping with 3-bit feedback outperforms the conventional STC without antenna grouping more than 1.8 dB at PER $= 10^{-2}$ (1.8 dB for 1/2 rate QPSK, 2.5 dB for 1/2 rate 16QAM, 2.4 dB for 1/2 rate 64QAM, and 3.2 dB for 2/3 rate 64QAM). As a higher MCS level is used, the performance gain is increased.

4.2.4.2 Codebook-Based Closed-Loop MIMO

The codebook words are employed in the feedback from mobile station (MS) to base station (BS). The MS learns the channel state information from downlink and selects a transmit beamforming matrix for the codebook. The index of the matrix in the codebook is then fed back to the BS. Each codebook corresponds to a combination of N_t, N_s, and N_i, which are the numbers of BS transmit antennas, available data streams, and bits for the feedback index, respectively. Once N_t, N_s, and N_i are determined in the MS, the MS will feed back the codebook index of N_i bits. After receiving an N_i bit index, the BS will look up the corresponding codebook and select the matrix (or vector) according to the index. The selected matrix will be used as the beamforming matrix in MIMO precoding.

4.3 Radio Resource Management

As the evolution of wireless networks allows for an increasingly wide range of services, the Radio Resource Management (RRM) function, which divides the available resources amongst competing applications, is receiving increased attention. There are several reasons that render RRM very important [49-75]:

1. RRM functions allow the support of a range of different requirements from the various services that the wireless network is required to support.

2. RRM may ensure the planned coverage (i.e., the area where the service is supported) for each service.

3. RRM may optimize capacity utilization.

Basically, the RRM has the complex task of maximizing the number of users that can be served satisfying their different service requirements, in time varying radio conditions and dynamic traffic behavior. In this section we will focus on the main RRM functions.

4.3.1   QoS Requirements

In order to guarantee a satisfactory end user quality, the transmission of a data flow, which is originated by the application, has to satisfy certain requirements that define the QoS profile for the information data stream of interest. Usually, the QoS attributes for a particular application/service are: required throughput, maximum acceptable delay, maximum acceptable delay jitter, and maximum acceptable bit error rate.

From the end user point of view, the following issues must be taken into account [53]:

- End users only care about the degree of QoS, and not about how it is provided.
- Only the QoS perceived by end user matters.
- The number of "user defined/user controlled" parameters has to be at a minimum.
- A derivation/definition of QoS attribute from the application requirements has to be simple.
- End-to-end QoS has to be provided.

A very frequent classification of the service class is based on the application/service's delay requirement. For example, in the UMTS standardization the following four service classes are defined: background class, interactive class, streaming class, and conversational class. The main difference between these classes is the delay sensitivity of the traffic. The conversational class is the most delay sensitive, while the background class is the most delay tolerant. Table 4.1 summarizes the UMTS QoS classes' main characteristics.

In the standardization the definition of QoS attributes are the same for GPRS Release 99 and UMTS. The QoS attributes for General Packet Radio Service (GPRS) Release 97/98 can be mapped on the Release 99 UMTS attributes as specified in [53]. However, the set of QoS attributes in UMTS is much larger than the set specified in GPRS Release 97/98 (Figure 4.4).

The throughput requirement is dependent on the information source. A throughput of 12 Kbps would be enough in order to transfer speech with GSM quality. For an audio stream with stereo quality, the throughput requirement is higher than 32 or 64 Kbps, while for a video communication it is higher than 128 Kbps.

The BER depends on the service class. For background (e-mail, file transfer protocol (FTP)) and interactive (web browsing) services, the received data should be error free. Due to the less stringent delay requirements of this service class, higher reliability can be achieved by error correction techniques (e.g.,

Table 4.1
UMTS QoS Classes

Traffic Class	Conversational	Streaming	Interactive	Background
Fundamental Characteristics	Preserve time relation (variation) between information entities of the stream Conversational pattern (stringent and low delay)	Preserve time relation (variation) between information entities of the stream	Request response pattern Preserve payload content	Destination is not expecting the data within a certain time Preserve payload content
Example of the Application	Voice, video games, voice telephony	Streaming video	Web browsing, network games	Background download of e-mail

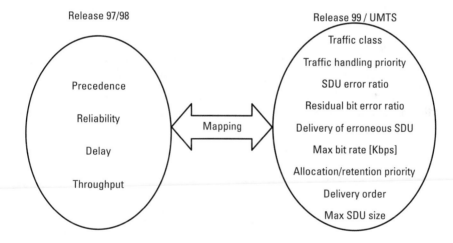

Figure 4.4 Mapping of QoS parameters.

packet retransmissions). The use of error correction mechanisms is rather limited by the delay requirement (e.g., for the interactive class). For the streaming and conversational classes, the acceptable BER depends on the type of information; for speech an acceptable BER is on the order of 10^{-2} or 10^{-3} and it may be even smaller for video transfer.

The application may specify its QoS requirements to the network by requesting a radio access bearer (RAB) with any of the specified traffic type, maximum transfer delay, delay variation, bit error rates, and data rates. In practice, it should be possible to define the main RAB characteristics from the service quality requirements:

- The transmission rate of RAB should be determined by the bandwidth requirement of the information source.
- The choice of dedicated or shared RAB should be based on the requirement for the maximum delay and delay jitter.
- The SNIR requirement, channel coding, and interleaving for the RAB should be based on the BER requirement.

4.3.2 General Formulation of the RRM Problem

Let us assume a cellular network with M mobiles in the service area and denote with $\mathbf{B} = \{1, 2, ..., B\}$ the set of all BSs used to provide the necessary coverage. Denote with C the number of available orthogonal channels in the system (i.e., the system capacity). The numbered set of all available channels is $\mathbf{C} = \{1, 2, ..., C\}$. The channels orthogonality could be established in different ways, such as in time and frequency domain in GSM or in the code domain in Wideband (W-CDMA). In GSM, as a representative of 2G systems, there is an intrinsic upper limit on the system capacity since the upper bound is the number of frequencies multiplied with eight time slots. On the other hand, in the WCDMA CDMA scheme, the set of orthogonal channels C is practically infinite and the capacity is determined by the interference condition in the system. A WCDMA system is, therefore, interference limited.

The link (power) gain matrix G characterizes the radio conditions in the system:

$$G = \begin{bmatrix} G_{11} & G_{12} & \cdots & G_{1M} \\ G_{21} & G_{22} & \cdots & G_{2M} \\ \cdots & & & \\ G_{B1} & G_{B2} & \cdots & G_{BM} \end{bmatrix} \tag{4.29}$$

The matrix element G_{ij} represents the link gain between the BS i and MS j; M represents the number of active mobiles. The gain matrix G is dynamic, the dimension M is changing, based on the offered load, and each element G_{ij} changes with the mobile movement. The radio resource management, taking into account the link gain matrix G, assigns [51]:

1. One or more access points from the set B;
2. A channel from the set C;

3. The transmit power of the BS and of the mobile.

The assignments 1 through 3 should maximize the number of users with a sufficient QoS. As outlined in the previous section, providing a stringent definition of the QoS for a communication service is a complex problem. A simple measure of the QoS-namely, the signal-to-interference + noise ratio (SNIR)-is here considered. This measure is strongly connected with the performance measures as the bit or frame error probability. Therefore, assignments 1 through 3 aim at maximizing the number of users for which the following inequality holds, for both the uplink (mobile-to-access port) and the downlink (access port-to-mobile):

$$SNIR_j = \frac{P_j G_{jj}}{\sum\limits_{\substack{M \\ m \neq j}} P_m G_{jm} \theta_{jm} + N} \geq \gamma_j \quad j = 1, \ldots, M \qquad (4.30)$$

In (4.30), $SNIR_j$ denotes the $SNIR$ at the receiver; P_j is the transmitter power used by the end user j; θ_{jm} is the normalized cross-correlation between the signal of interest and the interfering signal from the mobile m (other than j); N denotes the thermal noise power at the access port; and while γ_j is the target $SNIR$ of the service that is being used by the mobile j.

4.3.3 RRM in Future Wireless Systems

In this section, future RRM developments in wireless LANs and general RRM issues in mobile ad hoc networks (MANETs) are presented.

The Broadband Radio Access Network (BRAN) working group in ETSI is standardizing a wireless LAN for broadband radio access up to 54 Mbps. This new standard is called HIPERLAN/2 and includes physical layer and radio link control and data link control standards (see [61-65]). The interfaces toward other networks (e.g., UMTS) are made via specific design of convergence sublayers. The physical layer of HIPERLAN/2 is aligned with the physical layer of the IEEE 802.11a wireless LAN system [65]. Note here that wireless broadband networks based on the IEEE 802.11a standard became commercially available in 2002. Among the most important RRM functions in these WLAN systems are the link adaptation function and the radio resources allocation in the MAC frame.

In HIPERLAN/2 and IEEE 802.11 systems there are different pairs of modulation and coding schemes possible. Each pair results in different transmission rate and PER performance, depending on the radio channel quality (i.e., signal-to-interference ratio). The link adaptation scheme can dynamically change the pair modulation/coding scheme to optimize the throughput based on measured PER, signal level, packet size, and so forth. Extensive analysis of the physical layer performance of these two WLAN systems can be found in [63].

The MAC layer at the two WLAN standards is different. HIPERLAN/2 uses TDMA/TDD medium access with frame structure, as presented in Figure 4.5.

The allocation of radio resources is centrally scheduled by the access point (AP), which allows for implementation of scheduling algorithms, QoS differentiation, and resource reservation for services with stringent delay and delay variation requirements.

The duration of the broadcast channel (BCH) is fixed. Through these channels the relevant system information is conveyed to the terminals. The duration of the frame control channel (FCH), downlink (DL) and uplink (UL) phase, direct link (DiL) phase, and random channel (RCH) is dynamically adapted to the current load conditions of the AP. DiL phase is present if there are mobile terminals directly communicating on a peer-to-peer basis. Requests for the radio resources are signaled via the RCH, where contention for time slots is present. If scheduled data is transmitted, then FCH is present and it signals the frame structure. The transmission in the DL phase (from the AP to the terminals) and UL phase (from terminals to the AP) is contention free. The allocation of resources is signaled via the access feedback channel (ACH), as an answer to the resource request received via the RCH from the previous MAC frame.

The IEEE 802.11a standard, however, has a MAC scheme based on carrier sense multiple access with collision avoidance (CSMA/CA). A mobile terminal before transmission of data senses the radio channel. If the channel is free then the transmission can commence. Otherwise, an exponential back-off period is implemented before attempting the following packet transmission. This type of MAC (also known as distributed coordination function) makes the IEEE 802.11a standard more suitable for ad hoc wireless networks and non-real-time (background or interactive) type of applications. It should be mentioned that the standard also has contention-free MAC via the point coordination function (PCF), but this alternative is optional even though it could support real-time services. The advantage of the CSMA/CA is, however, the avoidance of a centralized scheduler that coordinates the radio transmissions.

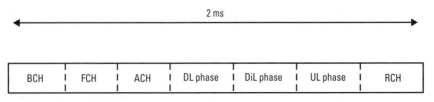

Figure 4.5 HIPERLAN/2 MAC frame structure.

Mobile ad hoc network, (MANETs) are also receiving attention in recent years. In Interent Engineering Task Force (IETF) there is a special working group (MANET working group) that investigates the routing protocols in ad hoc network. These networks are described by the IETF MANET group (source IETF MANET group):

A MANET is an autonomous system of mobile routers (and associated hosts) connected by wireless links, the union of which forms an arbitrary graph. The routers are free to move randomly and organize themselves arbitrarily; thus, the network's wireless topology may change rapidly and unpredictably. Such a network may operate in a stand-alone fashion, or may be connected to the larger Internet.

Currently, there are many research activities in the RRM field for MANETs. Due to the specific characteristics of MANETs, such as dynamic topology, limited node performance, distributed algorithms, and so forth, the development of RRM functions is a difficult task. The most important factors in MANETs are the coverage (or connectivity) and the capacity [66, 67]. An extensive field of research is QoS-aware routing (see [68] for one typical example) and MAC with QoS support [69]. A common MAC design for MANETs is driven by the hidden/exposed terminal problem. However, the QoS constraints for particular applications require from the MAC layer additional functionalities to provide certain guarantees and to make distinctions among different types of connections. Furthermore, MAC has to interact with QoS-aware routing in an appropriate way to provide the required communication quality over the whole path from source to destination (i.e., over the multiple hops). The current MAC proposals [69] that support QoS provisioning and differentiation are based on resource reservation along the connection's path, and differentiation between non-real-time and real-time connection establishment and maintenance.

The recent important development in the RRM field for multimedia wireless communication is that for systems beyond UMTS. In these systems the last wireless hop towards the end user could be carried over different radio access networks. Here, the interworking of UMTS, WLAN or HIPERLAN/2, and GSM/GPRS networks will play an important role. For example, the wireless operators could have in their coverage areas multiple possibilities for wireless communications via different radio access networks, as presented in Figure 4.6.

In this type of wireless networks, the setup of the Common RRM (CRRM) functions will play an important role in the efficient utilization of the available radio bandwidths. To achieve this goal the CRRM will perform traffic addressing towards less loaded wireless access systems. CRRM will choose the right radio access technology based on service requirements, current wireless system load, propagation conditions, interference, and capacity cost induced in the wireless network. This field is very challenging and interesting for future RRM research.

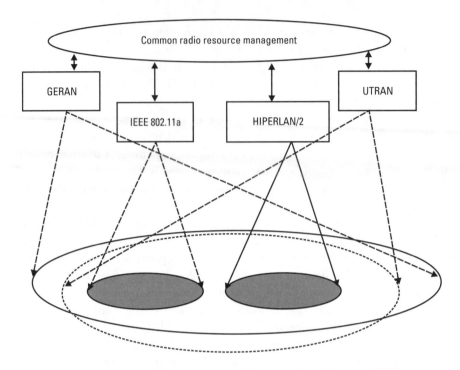

Figure 4.6 Interworking of different wireless access networks and common RRM.

4.4   Software Defined (SDR) Radio Communication Systems

As the number of wireless communication systems that users can use increase, there has been increasing demands for the coexistence of several mobile telecommunication services; for example, GSM, or IS-54, or IS-136, or IS-95, Japanese Personal Digital Cellular (PDC) or Personal Handy Phone System (PHS) or IMT-2000 system. Currently, if one wants to be globally connected, more than one terminal may be needed, though it is becoming more common to find multistandard terminals.

A person working on SDR is expected to be familiar with any communication system whether it is for radio transmission or multiple access purposes. Such schemes can be developed and written in programming languages like MATLAB or C. These computer simulation languages have a good relationship with software languages that configure digital signal processing hardware (DSPH) such as digital signal processors (DSP), field programmable gate arrays (FPGA), and application-specific integrated circuits (ASIC). A typical software language for DSPH is very high speed integrated circuit hardware description language (VHDL) and Verilog-HDL. DSPH has been utilized to configure the mobile terminals and base stations of mobile communication systems. DSPH is a general-purpose language, and the configuration can be programmed by downloading digital signal processing software (DSPS). This means that users can download DSPS describing the desired elemental components into the DSPH of only one terminal. This is the basic concept of an SDR communication system [76-90].

Software defined radio in theory will allow mobile terminal manufacturers to design and manufacture products that are independent of any particular specifications or standard. This means that users or the terminal itself can select the most appropriated air interface to be used based on channel conditions, traffic, cost, and so forth. There are benefits from the ecological point of view as well: SDR reduces the amount of hardware, which, in turn, reduces the amount of industrial waste.

Research into radio communication systems based on DSPH began in early 1990. These early studies, however, did not examine DSP-based radio communication systems from the viewpoint of reconfigurable radio communication systems. As far as we know, the first paper to coin the phrase "SDR" is that of Mitola [76]. Mitola described the fundamental and functional architecture of SDR. One prototype, the military SDR, SPEAKeasy, was introduced in [77]. SPEAKeasy can use several voice and data military services; the frequency band is 2 MHz to 2 GHz.

Most of the research towards realizing the ultimate communication system has focused on [78-82]:

1. The architecture of the DSPH;
2. The configuration of the analog signal processing hardware;
3. The method of downloading software to the hardware;
4. The method of application of software defined radio.

4.4.1 Definition of SDR Communication System

Figure 4.7 shows the basic configuration of the SDR of [76]. The radio consists of eight units: (1) antenna unit, (2) radio frequency signal processing unit (RFU), (3) intermediate frequency signal processing unit (IFU), (4) analog-to-digital conversion (ADC) and digital-to-analog conversion (DAC) unit, (5) baseband signal processing unit (BBU), (6) transmission control unit (TCU), (7) input/output (I/O) processing unit (IOU), and (8) end-to-end timing processing unit (TPU). Each unit will be described in detail.

1. Antenna unit. An omnidirectional, low loss, and broadband antenna is required because it can be used in a variety of wireless communication systems. Moreover, signal processing technology based on array

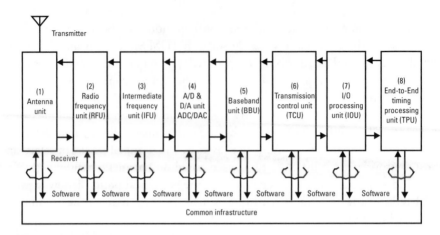

Figure 4.7 Basic configuration of the SDR in [76].

antennas makes it possible to select the performance of the SDR according to the surroundings and to perform optimum selection of the algorithm by using SDR technology. Such an antenna is called a smart antenna or software antenna [83, 84]. The software antenna is capable of space division multiple access (SDMA), in which the antenna steer the beam in the direction of selected users by computing appropriate weight coefficients for the antenna elements. Multiple access is achieved by changing the direction of the antenna or beams, or by interference cancellation, in which the antenna configures its direction to the desired user or allocates null points to the direction of undesired users or signals.

2. Radio frequency signal processing unit. In the transmitter's RFU, the signals coming from the IFU or BBU are up-converted to the radio frequency band signals, amplified, and transmitted to the antenna unit. At the receiver, the signals received by antenna unit are amplified to a constant level that is suitable for signal processing and directly down-converted to a lower frequency band such as IF band or baseband. The signal processing is done by an analog circuit. The linearity or efficiency of the RF amplifier and the conversion method to the lower frequency band at the receiver are the main discussion points.

3. Intermediate frequency signal processing unit. In this unit, the signals from ADC/DAC unit are up-converted to the IF band signal, amplified, and transferred to the RFU of the transmitter. At the receiver, the signals from the RFU unit are amplified to an adequate level for signal processing in the IFU and directly down-converted to a suitable frequency for the ADC/DAC unit or baseband unit. When the signals of

several systems are received at the receiver, the required frequency band must be selected by using a filter.

4. Analog-to-digital conversion and digital-to-analog conversion unit. In this unit, the digital signal from the baseband unit is converted to an analog signal by using a DAC and the converted signal is transferred to the upper frequency band unit (IFU or RFU). At the receiver, the signals from the IFU or RFU are amplified to an adequate level for analog-to-digital conversion. The stabilized signal is then sampled by ADC and converted to a digital signal. In this unit, the sampling method is a key technology.

5. Baseband signal processing unit. In this unit, data is digitally modulated and transferred to the ADC/DAC unit of the transmitter. Transmitted data is recovered by using the sampled signal from the ADC/DAC unit and digital signal processing at the receiver. The basic configuration of the BBU is shown in Figure 4.8 [77]. In the BBU of the transmitter, frame, coding, mapping and modulation, and transmission filter blocks are the key blocks. On the other hand, in the BBU of the receiver, receiver filter, code and symbol timing, sampling rate conversion (resample), demapping and demodulation, and decoding blocks are the key blocks. Moreover, in the BBU of the receiver, the fading compensation (equalization) block and the interference cancellation block for eliminating undesired signals are present. In most cases, the BBU is configured by several DSPH such as DSP and/or FPGA and/or ASIC. All component blocks are described using DSPS written in VHDL or Verilog-HDL and compiled. The BBU's configuration can be modified by changing the DSPS.

6. Transmission control unit. In this unit, the input bit stream format for the BBU is configured at the transmitter by adjusting the transmission protocol of the MAC layer, and at the receiver, the detected data from the BBU is checked according to the data format of the transmission protocol of the MAC layer. If the number of bit errors in the detected data is large, retransmission is requested. In addition to this transmission control, this unit can manage cryptograph. In most cases, TCU can be configured by a range of DSPHs, and all the component blocks are also described using DSPS. By changing DSPS, the TCU can configure the transmission protocol as needed.

7. Input/output processing unit. In the mobile station, all information data comes from a handset, PDA terminal, or personal computer, and all received data comes back to these terminals or computers. The I/O and timing are managed so as to connect with the external terminals flexibly.

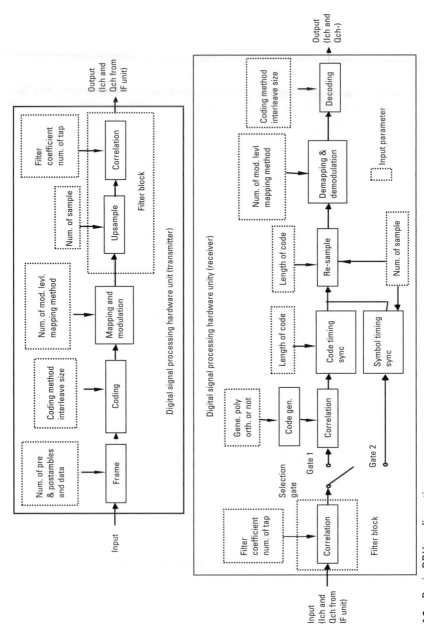

Figure 4.8 Basic BBU configuration.

8. End-to-end timing processing unit. This unit controls the transmission delay between transmitter and receiver. For example, the transmission delay for voice must be shorter than 150 ms (typically).

In most SDR systems, several software programs, which describe all telecommunication components in DSPS language, are used to configure the components on the DSPH. This software can be readily changed to suit the requirements of a particular system. Such a SDR system is called a full-download-type SDR system. Figure 4.9 shows the configuration of a full-download-type SDR system. The system has an RFU, IFU, and BBU as in the basic system. Moreover, it has a TX module, which is related to the transmitter, and an RX module, which is related to the receiver.

Implementing a specific telecommunication system with a full-download-type SDR system requires that all necessary DSPS be downloaded to the BBUs before starting communication. DSPS blocks, including frame block, encoder block, mapping and modulation block, and filter block (Figure 4.8) are downloaded to the BBU of the TX module. DSPS blocks, including filter block, equalizer block, detector and decoder block (Figure 4.9), are also described in DSPS and downloaded to the BBU of the RX module. After software has been downloaded, the configuration check program is executed. The BBU then

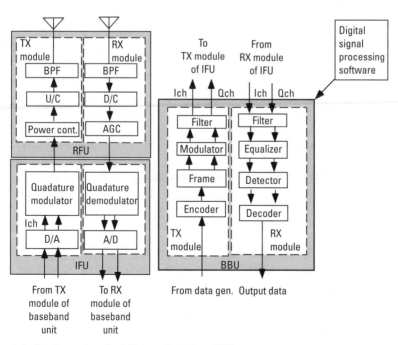

Figure 4.9 Configuration of a full-download-type SDR system.

configures the required baseband modulation and demodulation circuit. Then, transmitted data can then be fed into the BBU of the TX module.

In the BBU of TX module, this data are formatted into frames, modulated and converted into two signals: in-phase channel (Ich) and quadrature-phase channel (Qch) signals by the DSP blocks mentioned above. These signals are fed into the IFU of the TX module.

In the IFU of the TX module, the digitally modulated Ich and Qch signals are converted from digital to analog by a D/A converter block. These signals are quadrature modulated on the IF band and sent to the RFU of the TX module. In the RFU of the TX module, the quadrature-modulated signal is up-converted to the RF band by power control part before being transmitted.

To receive the RF signal, the received signal is fed into the RFU of the RX module. Here, the received RF data is bandpass-filtered, which eliminates spurious signals, and down-converted to the IF band. The automatic gain control (AGC) block keeps the power of the down-converted signal at a constant level. This power-controlled signal is fed into the IFU of the RX module.

In the IFU of the RX module, the received signal from the RFU is split into Ich and Qch signals by using a quadrature demodulator block. The A/D converter block then oversamples and transfers them into the BBU of the RX module.

In the BBU of the RX module, all telecommunication component blocks have been implemented into the DSPH before starting communication, and the configuration of what has been checked by a test program. The oversampled Ich and Qch signals are filtered and equalized by a customized method, and they are detected and decoded by using the filter, equalizer, and decoder blocks in the BBU of RX module.

4.4.2   Advantages of SDR Communication Systems

In full-download-type SDR, the system configuration can be changed on demand. There are many advantages not only for operators and service providers but also for government and commercial customers.

In particular, for commercial operators and service providers the advantages are:

1. Global roaming services;
2. Upgradeable terminals;
3. New services added without having to change the terminal;
4. Bugs fixed without the need to recall the product;
5. Versatile software (i.e., wireless communication software that can be installed in other electrical products as well as in the mobile terminal).

For government agencies, the advantages are:

1. Global roaming services can be offered to customers;
2. SDR reduces the variety of hardware, since several standards can be implemented in a single mobile terminal.

For customers, the advantages are:

1. Unlimited global roaming;
2. One terminal for many services;
3. New services provided without needing to upgrade hardware;
4. New services added to the terminal without changing the terminal;
5. Bugs fixed without needing to recall the product.

4.4.3 Problems in SDR Communication Systems

As shown in Section 4.4.2, SDR has many advantages, but the technology must overcome the following problems.

1. The volume of software downloaded to the DSPH increases as the contents of the required telecommunication component blocks become more complicated. As a result, the download time is lengthened. In addition, the software files to be transmitted need to be protected by an adequate channel coding scheme to make the transmission more robust to fading and interference. In this case, the download time becomes even longer [87, 88].
2. The period for the configuration check of the DSPH also increases since the contents of the required telecommunication component blocks become more complicated. The problem also affects the stability of the operating characteristic of the DSPH when there is not a sufficient period of time for the configuration check [87, 88].
3. In the downloaded software, there are often several component blocks containing manufacturer-specific know-how (e.g., the optimization method or calculation algorithm for some special blocks). There is the possibility that this know-how may leak out when the software is downloaded. There is also the possibility that it may be tampered with [87, 88].

In addition to the above problems, if SDR controls RF and IFU as well as the baseband unit, the following issues must be considered.

1. By using SDR, a user must download only the software describing the elemental components to the hardware for realizing a particular communication system. However, if several communication systems operating on different frequency bands are integrated into one mobile communication hardware by SDR technology, several antenna units, RFUs, IFUs, and ADC/DAC units are still needed in the mobile communication hardware. Moreover, a user may want to use several systems simultaneously [89], in which case several ADC/DAC units, baseband units, transmission control units, I/O processing units, and end-to-end timing processing units must be prepared. The method of managing several units must be considered.

2. If several communication systems operating on different frequency bands are integrated into one mobile communication hardware by SDR technology, a broadband antenna must be used. Since the bandwidth of antenna is limited, however, several antennas are needed. The placement of these antennas in the terminal is an important design issue to be taken into consideration.

Moreover, from the viewpoint of the entire system, the following questions arise:

1. Which services are to be integrated in one SDR terminal? Say that a multimode terminal can easily be realized by developing a specific ASIC for each communication system and planning these ASICs on one circuit board. In this case, SDR technology is unnecessary. Therefore, SDR technology should be targeted to application fields in which many communication systems are required.

2. How and in which field is the SDR technology socially recognized? There are many applications for SDR technology. These applications must be identified.

3. How and in which field is standardization to be done? How should the software be protected against viruses or hackers? In particular, SDR's capability of international roaming by changing software instead of changing terminal becomes a serious threat, if a virus is included that can reconfigure a wireless communication system. A simple example is the reconfigured terminal in which the transmitter no longer matches the receiver. Moreover, a virus altering the transmission power to a much higher level than the regulation threatens to shut down all wireless communication systems in the world. Reference [90] describes wireless terrorism. In wireless terrorism, all users of SDR effectively become potential terrorists. To prevent such an

incident from occurring, standardization must be done. In addition, the organizations and radio that certify SDR terminals must be developed. Software radio thus presents a new paradigm in radio law.

4.4.4 Future Applications of SDR Communication Systems

4.4.4.1 Future Telecommunication Applications

The following applications are envisioned for the future.

Mobile Communication Terminal. A SDR technology-based mobile communication terminal allows users to select a system by changing the software of the DSPH. Moreover, users can select the provider company that they like. However, to implement them will require miniaturization, reduction of the power consumption and cost of the hardware, as well as reduction in software download time. An efficient system selection procedure is also required.

Mobile Communication Base Station. The base station that has SDR technology would make it easy to implement new communication systems and fix bugs. In addition, a base station could use more than one algorithm to fight against fading. Moreover, by using an adaptive array antenna, the antenna beam shape could also be controlled with software. To realize this application, we must make a rule for programming software that configures components for the mobile communication system to change the program easily and to make the connection with other components. Moreover, a fast ADC or digital signal processor is needed to perform real-time transmission.

Broadband Radio Access System. Software radio technology can enhance the flexibility of a broadband radio access system, which operates between base stations and buildings. For example, software can be used for components such as distortion compensators and interference suppressors. The modulation scheme also can be changed, and the best components selected. Using an adaptive array antenna would allow the beam shape to be controlled with software. To do so, each module should be implemented by software, and also, the connections between modules should be programmable by software. A high speed ADC or digital signal processor must also be used to perform real-time transmission.

Intelligent Transport System. The communication systems utilized in the intelligent transport system (ITS) are shown in Table 4.2. They fall in three categories: communication-based systems, control-based systems, and broadcasting-based systems. Representative examples of communication-based systems are Personal Digital Cellular (PDC), PHS, and digital cellular using CDMA.

Table 4.2
Simulation Models

System	PHS	GPS	ETC
Frequency	1.9-GHz band	1.5-GHz band	5.8 GHz
Modulation	p/4DQPSK TDMA-TDD	BPSK + DS-Spread Spectrum	ASK (Manchester code)
Data Rate	384 Kbps	50 bps	1,024 Kbps
Bandwidth	300 kHz/1ch	1.023 MHz	Less than 8 MHz

The global positioning system (GPS) system, vehicle information and communication system (VICS) system, and radar are representative examples of control-based systems. The Advanced Cruise-Assist Highway System (AHS)-which provides information on traffic accidents, road obstructions, and meteorological phenomenon-and the electronic toll collection (ETC) system will be mounted in cars. Representative examples of broadcasting-based systems are radio broadcasting systems, analog-television broadcasting systems, satellite broadcasting systems, and digital television systems. As mentioned above, the number of communication systems appearing in the ITS system will likely increase. This means that space-saving and integration of system components must be chief concerns if so many services are to be provided in the limited space of a car. SDR technology envisions that these services will be provided by only one small terminal. Figure 4.10 shows the potential application. The method of introducing SDR technology is the same as described in the first three applications above. In the ITS, the number of services that can be used in a car is more than 10. Therefore, we urge implementation of SDR technology as soon as possible.

4.4.4.2 Broadcasting Applications

Software radio technology can realize automatic switching between terrestrial wave/BS/CS/CATV. It is easy to introduce new services on existing hardware because only the software needs to be changed, as was discussed before. Moreover, receivers compatible with the standards of many countries or regions can be realized. This would have an impact on the promotion and distribution of international products. Such a multimode, multistandard receiver must have a broadband antenna and software describing the components of the particular communication system, and this means that the connectivity between components must be increased.

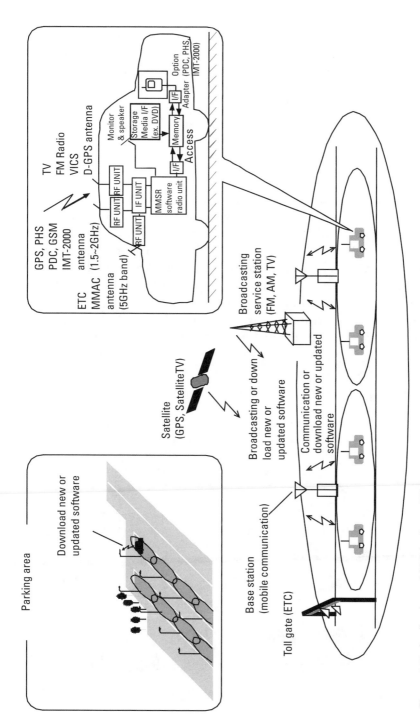

Figure 4.10 Applications of SDR for future ITS system.

4.4.4.3   Private Networks

Nowadays, many private networks exist for diverse purposes, like education, office, community, ITS, emergency, and commercial use (Figure 4.11). Current terminals for these networks do not have any connectivity with each other. Software radio can be used to create a universal terminal that works on any private network. In the near future, these huge networks will not only complement but also compete with conventional public networks.

4.4.4.4   Certification Method of SDR

One problem facing practical implementation of SDR is wireless terrorism. The potential types of damage are shown in Figure 4.12. In some cases, downloaded data may be illegally altered or have embedded in it a computer virus. Altered or virus software may cause unexpected and even dangerous changes in the base station (e.g., the BS could be commanded to shut down or increase transmission power over permitted levels, or it could be permanently damaged). In order to avoid such wireless terrorism, a new certification method for the systems using SDR must be required in the certification organizations.

An example of certification is shown in Figure 4.13. The certification method has two stages. In the first stage, the certification organization checks the relationship between input data and output data by changing the software that configures several systems. In this case, for each system, the transmission power, transmission bandwidth, and electric power leakage in the adjacent

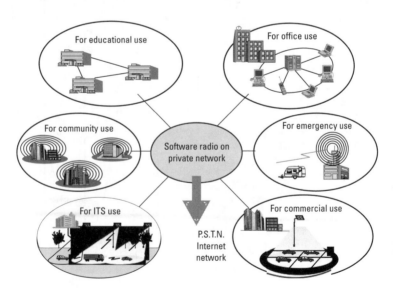

Figure 4.11 Private network applications of SDR.

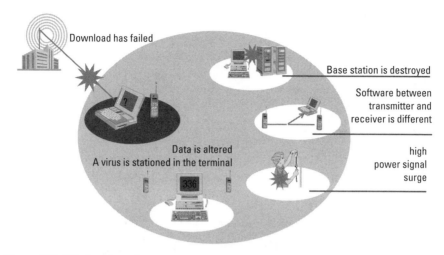

Figure 4.12 Wireless terrorism.

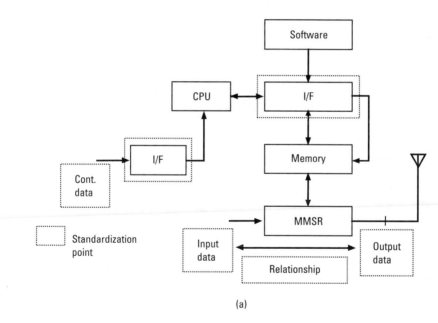

(a)

Figure 4.13 Certification method of SDR equipments: (a) first phase; and (b) second phase.

frequency band are measured and evaluated by comparing the radio law of each country.

If the integrity of the software is verified and the new reconfiguration is acceptable, the certification organization gives a certification password to the software. The certification password is used in the handshake between the

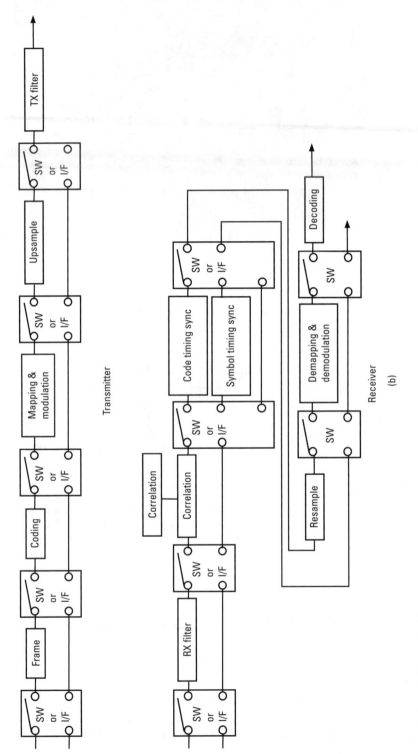

Figure 4.13 continued.

DSPH and memory before downloading the software to the DSPH. During the handshake, if the password is certificated, the software is downloaded from memory to the DSPH, otherwise the SDR equipment issues a warning. The public key cryptography technique is a candidate for the certification password. Moreover, the download protocol between the memory and the DSPH must be standardized and the protocol must be certificated.

The second stage of certification involves the modules that configure SDR. To certify each component, the configuration of SDR must be changed. An example of configuration is shown in Figure 4.13(b). Here, switching or interface modules are inserted between functional modules to configure the SDR equipment. These switching or interface modules can be controlled from outside of the DSPH. By switching modules, each functional module is certificated. The switching or interface module must be standardized for such a certification procedure to work.

4.4.5 Summary

One of the future applications of the software programming method is in the SDR communication systems. The concept of SDR bridges the software programming and hardware implementation.

This section has described the SDR communication system, its advantages and its problems, as well as some possible future applications. This technology will likely become key to the success of forth generation mobile communication systems, because nowadays there are many systems in the world and these systems must be integrated.

4.5 IP Network Issues

The evolution towards a common, flexible, and seamless IP-based core network that will connect heterogeneous networks gives rise to several issues at the network layer [91-160]. The need for ensured QoS is the key of this evolution. Fundamentally, the day that packet-switched networks can credibly approach the QoS of circuit-switched networks is the day customers stop paying for two networks. The QoS problem involves integrating delay-sensitive applications such as voice, audio, and video onto a single network with delay-insensitive applications, such as e-mail, fax, and static file transfer. That network must be able to discriminate, differentiate, and deliver communications, content, and commerce services. Moreover, it needs to support communication services creation, modification, bundling, and billing in a way that is unobtrusive yet powerful for end users, especially business users. The scenario becomes more challenging as QoS and security issues are faced in a mobile environment. The

Internet has not been designed with mobility in mind and lacks mechanisms to support mobile users. Some architectures have been proposed for supporting mobility in the Internet. The most important is Mobile IP, which will be discussed in Section 4.5.2. Before that, some guidelines behind mobility management are discussed in Section 4.5.1. Proposals to solve some of the open issues in Mobile IP are described in Section 4.5.3.

4.5.1   Mobility Management

When a mobile node is roaming through one or more service areas, mobility management mechanisms may be required to locate it for call delivery and maintenance of its connections. Generally, in cellular systems, mobility management is performed through two main mechanisms:

1. *Location management* is used for discovering the current attachment point of the mobile user for call delivery. It consists of two phases. In the first phase, called the location *registration* (or *location update*), the mobile terminal periodically notifies the network for its new access point, allowing the network to authenticate the user and revise the user's location profile. The second phase is *call delivery*. The network is queried for user location profile and the current position of the mobile host is found.

2. *Handoff management* enables the user to keep its connection alive as it moves and changes its access point to the network. This is performed in three steps: *initiation, connection generation,* and *data flow* control.

When the user moves within a service area or cell and changes communications channels allocated by the same BS, the mobility management procedure is called *intracell* handoff. *Intercell* handoff occurs when the user moves into an adjacent service area or cell for which all mobile connections are transferred to a new BS. While performing handoff, a mobile terminal can be simultaneously connected to two or more BSs and use some kind of signaling diversity to combine the multiple signals. This condition is called *soft handoff*. In *hard* handoff, the mobile device switches from one base station to another with active data being forwarded on only one path at a time.

4.5.1.1   Mobility Classes

In general, there are four categories that support IP mobility:

1. Pico-mobility is the movement of a mobile node (MN) within the same BS. The operating space is the space around person that typically

extends up to 10m in all directions and envelops the person, being either stationary or in motion.

2. Micro-mobility is the movement of an MN within or across different BSs within a subnet; this occurs very rapidly. Management of micro-mobility is accomplished using link-layer support (layer 2 protocol), which is already implemented in existing cellular networks.

3. Macro-mobility is the movement of an MN across different subnets within a single domain or region; this occurs relatively less frequently. This is currently handled by Internet mobility protocols such as Mobile IP.

4. Global mobility is the movement of an MN among different administrative domains or geographical regions. This is also handled by layer 3 techniques such as Mobile IP.

Mobility management has a responsibility of providing uninterruptible connectivity during micro- and macro-mobility, which usually occur over relatively short time scales. Global mobility, on the other hand, usually involves longer time scales. Therefore, the goal is just to ensure that mobile users can reestablish communication after they change the domain, but not necessarily to provide uninterruptible connectivity.

4.5.1.2 Architectures for Mobility Supporting

There are several frameworks that support mobile users, and the IETF standardizes two of them: Mobile IP and Session Initiation Protocol (SIP).

1. Mobile IP [91] supports application-layer transparent IP mobility. The basic Mobile IP protocol does not require protocol upgrades in stationary correspondent nodes (CNs) and routers. Its drawback is that it does not consider the integration of additional functions such as authentication and billing, which are critical for successful adoption in commercial networks.

2. SIP [92] is an application-layer control (signaling) protocol that can establish, modify, and terminate multimedia sessions or calls. The main disadvantage of SIP mobility is that it cannot support transmission control protocol (TCP) connections and is also not an appropriate solution for micro- or macro-mobility.

SIP mobility will not be a subject of further discussion in this chapter; instead the focus will be on Mobile IP.

4.5.2   Mobile IP

Mobile IP refers to a set of protocols, developed and still under development by the IETF to allow the Internet Protocol to support the mobility of a node [91]. The idea for Mobile IP first emerged in 1995 and since then it has undergone some changes.

First of all, it is necessary to understand what makes the IP mobility complicated. An IP address consists of two parts: a prefix that identifies the subnet in which the node is located, and a part that identifies the node within the subnet. Routers use look-up tables to forward incoming packets according to their destination addresses. A router does not store the addresses of all computers in the Internet, which is not feasible. Only prefixes are stored in the routing tables and some optimizations are applied. Therefore, as a receiver moves outside the original subnet, it can no longer be reached. A new IP address should be assigned to the mobile node, but this operation requires time. Specifically, the Domain Name System (DNS) needs time to update its internal tables for mapping a logical name to an IP address. Therefore, this approach cannot work if the node moves quite often. Moreover, a browser and a Web server being, respectively, client and server in communicating, generally use the TCP/IP protocol suite to establish a reliable end-to-end communication. Using TCP means that a virtual connection must be established before data transmission and reception. When a logical circuit is created, both connection sides must be assigned port numbers in order to let application layers keep track of the communications. Every end-to-end TCP connection is identified by four values: client IP address, client TCP port, server IP address, and server TCP port. These values must be constant during the conversation. Therefore, a TCP connection will not survive any address change.

Allowing the mobile nodes the use of two IP addresses provides the solution to the above problem. In Mobile IP, *home address* remains unchanged regardless of where the node is attached to the Internet and it is used to identify TCP/IP connections. Another IP address, which is dynamic, is assigned to the mobile node when it moves to another network different than its home network. This address is called the *care-of address* (COA) and it is used to identify the mobile node point of attachment in the network topology. Using its home address, the mobile node is seen by other Internet hosts as a part of its own home network. In other words, other Internet hosts do not have to know the actual mobile node's location. The *home agent* (HA) is located in the home network. Whenever the mobile node is not attached to its home network but to another network, called the *foreign network* (FN), the home agent gets all packets destined for the mobile node and arranges their delivering to the *foreign agent* (FA). The FA then sends the packets to the mobile node. An HA can be implemented on a router that is responsible for the home network. This position is quite appropriate because without any optimization to mobile IP, all packets for the MN have to go through the router anyway. An alternative solution consists in making the router behave as a manager for MNs belonging to a virtual home network. In this case, all MNs are always in a foreign network. The HA could also be implemented on an arbitrary node in the subnet. This solution is necessary when the router software cannot be changed, but it has the disadvantage that there is a double-crossing of the router by the packet if the MN is in a foreign network.

In general, the Mobile IP concept can be considered as a combination of three major functions:

1. *Discovery mechanism*, which allows mobile computers to determine their IP address as they move from network to network;
2. *Registering* of the new IP address with its home agent;
3. *Tunneling* (delivery) of packets to the new IP address of the mobile node.

4.5.3   Evolution of Mobile IP

Mobile IPv4 is insufficient to meet the requirements of the evolving communication scenario for several reasons.

Macro-Mobility.   Open issues in Mobile IPv4 related to macro-mobility management include the following:

- *Asymmetric (Triangle) routing.*
- *Inefficient direct routing:* The routing procedure in Mobile IPv4, with respect to the number of hops or end-to-end delay, is inefficient.
- *Inefficient home agent notification:* When a mobile node hands off from one network to another, it has to notify the home agent about that. This operation is inefficient in Mobile IPv4.
- *Inefficient binding deregistration:* In Mobile IPv4, if a mobile node moves to a new FA, then the previous FA could not release the resources occupied by the mobile node. The previous FA must wait until a binding registration lifetime expires.

Micro-Mobility.   Mobile IPv4 is not concerned with micro-mobility issues. Since it is expected that in the near future, Mobile IP will be a core wireless IP architecture interconnecting different wireless networks, the interoperability between macro-mobility and micro-mobility issues is very important. There are several open issues in this context:

- *Local management of micro-mobility events:* In micro-mobility dimension, handoffs occur very frequently. Therefore the handoff procedures should be managed as locally as possible.

- *Seamless intradomain handoff:* After intradomain handoff, the IP data stored into the previous entities (e.g., base stations) should be transferred to the new BS.

- *Mobility router crossings in an intranet:* During intradomain handoff, the router crossings should be avoided as much as possible.

4.5.3.1   Mobile IPv6

The IETF began work in 1994 on IPv6, a proposed standard designed to address various problems with Mobile IPv4. The IETF skipped the IPv5 standard because network operators had already adopted many of the protocol's potential provisions before it could be adopted.

Mobile IPv6 is supposed to replace Mobile IPv4. One of the main advantages of IPv6 over IPv4 is the higher number of IP addresses. IPv4's 32-bit addressing scheme can support a theoretical maximum of 4.29 billion IP addresses. However, due to the operational inefficiency, useful IP addresses are about 200 million. IPv6 offers a 128-bit addressing scheme that permits about $2^{1,033}$ useful IP addresses.

Other major differences between Mobile IPv4 and the Mobile IPv6 [146] are:

- Mobile IPv6 supports the mechanism of route optimization. This feature is already an integral part of the Mobile IPv6 protocol. In Mobile IPv4 the route optimization feature is just a set of extensions that may not be supported by all IP nodes.

- Mobile IPv6 specifies a new feature that allows mobile nodes and Mobile IP to coexist efficiently with routers that perform ingress filtering [147]. The packets sent by a mobile node can pass normally through ingress filtering firewalls. This is possible because the COA is used as the source address in each packet's IP header. Also, the mobile node home address is carried in the packet in a *home address destination option* [146]. In this way, the use of the COA in the packet is transparent above the IP layer.

- The use of the COA as the source address in each packet's IP header simplifies the routing of multicast packets sent by a mobile node. Thus, the tunneling of the multicast packets to its home agent will no longer be necessary in Mobile IPv6. The home address can still be used and it is compatible with multicast routing that is based in part on the packet source address.

- Neighbor discovery [148] and address autoconfiguration [149] enable the functionality of the foreign agents instead of using special routers. The foreign agents are not required any more in Mobile IPv6. Thus, the issue of mobility router crossings in an intranet is resolved.

- The Mobile IPv6 uses IPsec for all security requirements (e.g., sender authentication, data integrity protection, and replay protection for binding updates-which serve the role of both registration and route optimization in Mobile IPv4). Mobile IPv4 is based on its own security mechanisms for each function, based on statically configured mobility security associations.

- Mobile IPv6 provides a mechanism for supporting bidirectional (i.e., packets that the router sends reach the mobile node, and packets that the mobile node sends reach the router) confirmation of a mobile node ability to communicate with its default router in its current location. This bidirectional confirmation can be used to detect the black hole situation, where the link to the router does not work equally well in both directions. Unlike Mobile IPv6, Mobile IPv4 does not support bidirectional confirmation. Only the forward direction (packets from the router are reaching the mobile node) is confirmed, and therefore the black hole situation may not be detected.

- In Mobile IPv6, the correspondent node sends packets to a mobile node while it is away from its home network using an IPv6 routing header rather than IP encapsulation, whereas Mobile IPv4 must use encapsulation for all packets. In this way ReSerVation Protocol (RSVP) operation in Mobile IP is enabled and also the problem of ingress filtering is partially solved. In Mobile IPv6, however, the home agents are allowed to use encapsulation and tunnel the packets to the mobile node.

- In Mobile IPv6, the home agent intercepts the packets, which arrive at the home network and are destined for a mobile node that is away from home, using IPv6 neighbor discovery [148] rather than Address Resolution Protocol (ARP) [100] like in Mobile IPv4.

- IPv6 encapsulation (and the routing header) removes the need to manage tunnel soft state, which was required in Mobile IPv4 due to limitations in ICMP error procedure. In Mobile IPv4, an ICMP error message that is created because of a failure to deliver an IP packet to the COA will be returned to the home network, but it will not contain the

IP address of the original source of the tunneled IP packet. This is solved in the home agent by storing the tunneling information (i.e., which IP packets have been tunneled to which COA), called *tunneling soft state.*

- Mobile IPv6 defines a new procedure, called *anycast.* Using this feature, the dynamic home agent discovery mechanism returns one single reply to the mobile node, rather than the corresponding Mobile IPv4 mechanism that used IPv4 directed broadcast and returned a separate reply from each home agent on the home network. The Mobile IPv6 mechanism is more efficient and more reliable. In this way, only one packet requires to be replied to the mobile node.

- In Mobile IPv6, an advertisement interval option on router advertisements (equivalent to agent advertisements in Mobile IPv4) is defined. This allows a mobile node to decide for itself how many router advertisements (agent advertisements) it is ready to miss before declaring its current router unreachable.

- The IPv6 destination options permit all Mobile IPv6 control traffic to be piggybacked on any existing IPv6 packets. Mobile IPv4 and its route optimization extensions require separate UDP packets for each control message.

4.5.3.2 Macro/Micro-Mobility Extensions to Mobile IP

Several protocols and frameworks have been proposed to extend Mobile IP to better support micro- and macro-mobility in next generation wireless cellular environments.

Fast and Scalable Handoffs for Wireless Internetworks

This is an extension to Mobile IP [150] that uses hierarchical FAs to handle macro-mobility. This architecture assumes BSs to be network routers; for that reason, it is not compatible with current cellular architectures, where BSs are simply layer 2 forwarding devices. Moreover, deploying a hierarchy of FAs imposes complex operational and security issues (especially in a commercial multiprovider environment) and requires multiple layers of packet processing over the data transport path. The presence of multiple layers of mobility-supporting agents also significantly increases the possibility of communication failure, since it does not exploit the inherent robustness of Internet routing protocols.

Mobile-IP Local Registration with Hierarchical Foreign Agents

This protocol is an IETF proposal [151], and it solves the issues of triangle routing and inefficient direct routing. The basic mechanism of this architecture deploys hierarchical FAs for seamless mobility within a domain. During the COA discovery procedure multiple FAs are advertised using the agent advertisement message. The COA registration will be provided for the FA that is the lowest common FA ancestor at the two points of attachment of interest. The requirement for hierarchical agents in Internet mobility architecture remains an open issue. Even though it does not appear to be a critical consideration in the immediate future, it is possible that hierarchical mobility management will become more attractive as the IP security infrastructure matures and deployment of mobile multimedia terminals gets much larger.

TeleMIP

TeleMIP stands for Telecommunication Enhanced Mobile IP [152]. It achieves smaller handoff latency by localizing the scope of most location update messages within an administrative domain or a geographical region. However, this architecture faces the problem of inefficient home agent notification. TeleMIP introduces a new logical entity, called the *mobility agent* (MA), which provides a mobile node with a stable point of attachment in a foreign network. While the MA is functionally similar to conventional FAs, it is located at a higher level in the network hierarchy than the subnet-specific FAs. Location updates for intradomain mobility are localized only up to the MA. Global location updates are necessary only when the mobile changes the administrative domains. The TeleMIP allows efficient use of public address space, by permitting the use of private addresses for handling macro-mobility. Reduction of the frequency of global update messages overcomes several drawbacks of existing protocols-such as large latencies in location updates, higher probability of loss of binding update messages, and loss of in-flight packets-and thus provides better mobility support for real-time services and applications. The dynamic creation of mobility agents (in TeleMIP) permits the use of load balancing schemes for the efficient management of network resources. Its drawback is the potential nonoptimal routing within the domain.

Wireless IP Network Architecture by TR45.6

Another framework for IP-based mobility management was developed by the Telecommunications Industry Association (TIA) Standards Subcommittee TR45.6 [153] to target 3G cellular wireless systems. This architecture is consistent with the requirements set by the ITU for IMT-2000. Therein, solutions are provided to the issue of inefficient direct routing. The framework uses Mobile IP with fast rerouting for global mobility. For macro-mobility, the scheme proposes the use of dynamic HAs (DHAs), which reside in the serving network and are dynamically assigned by the visited authentication, authorization, and accouting server. The DHA allows the roaming user to gain service with a local access service provider while avoiding unnecessarily long routing. The architecture defines a new node called a *packet data-serving node* (PDSN) (which contains the FA), and uses visitor location register(VLR)/home location register (HLR) (ANSI-41 or GSM-MAP) authentication and authorization information for the access network. The mobile node is identified by a network access identifier (NAI) [154] in the visiting or foreign network. Within the registration process, an MN sends the registration message to the FA, which in turn interacts with an authentication, authorization, and accouting server residing in that network or uses the broker network for authentication with the home network.

Micro-Mobility Extensions to Mobile IP

Due to the fact that the basic Mobile IP protocol [91] is only concerned with the macro-mobility management, some other solutions are required to enhance the Mobile IP functionality to support micro-mobility. The following overview shows the results of several current research activities in this area.

Wireless Network Extension Using Mobile IP.   This micro-mobility management framework [155] is combined with Mobile IP. It provides solutions to the issue of local management of micro-mobility events. The development of this scheme is realized in the Motorola iDEN architecture. Micro-mobility events should be managed more efficiently than macro-mobility events, because they can happen with relatively high frequency. Therefore, the procedures and participants are being kept as local as possible. The micro-mobility procedures are managed by a data gateway, thus achieving the previous condition. The macro-mobility between iDEN subnetworks and other subnetworks is accomplished by implementing Mobile IPv4 in the FA and HA (Figure 4.14).

HAWAII.   The Handoff-Aware Wireless Access Internet Infrastructure (HAWAII) [156] proposes a method for using a separate binding protocol to handle micro and macro-mobility. For global mobility, it uses Mobile IP. Using this architecture, a solution to the issue of local management of micro-mobility events is provided. It uses a two-layer hierarchy for mobility management. When the MN moves into a foreign domain, it is assigned a collocated COA from that domain, and the MN retains its COA unchanged while moving within the foreign domain. Thus, the movement of the MN within a domain is transparent to the HA. This protocol uses path setup messages to establish and update host-based routing entries for MNs in some specific routers within the domain; other routers not in the path are kept in the dark about the MN's new COA. When a CN sends packets to a roaming user, it uses the MN home IP address. The HA intercepts the packets and sends the encapsulated packet to the MN's current border router. The border or root router decapsulates and again encapsulates the packet to forward it to either the intermediate router or BS, which decapsulates the packet and finally delivers it to the MN (Figure 4.15).

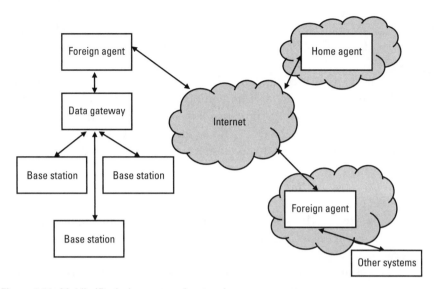

Figure 4.14 Mobile IP wireless network extension.

Cellular IP. Cellular IP [157, 158] proposes an alternative method for providing mobility and handoff support in a cellular network, which consists of interconnected cellular IP nodes. Solutions to issues like local management of micro-mobility events and seamless intradomain handover are provided. This

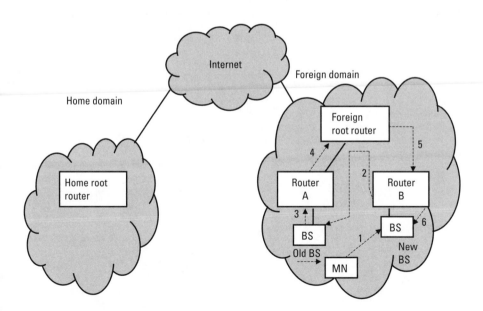

Figure 4.15 HAWAII architecture.

protocol uses Mobile IP for global mobility. It is very similar to the host-based routing paradigm of HAWAII. Specifically, Cellular IP provides local mobility (e.g., between BSs in a cellular network) (Figure 4.16).

The architecture uses the home IP address as a unique node identifier, since MN addresses have no location significance inside a cellular IP network. When an MN enters a Cellular IP network, it communicates the local gateway (GW) address to its HA as the COA. Nodes outside the Cellular IP network do not require any enhancements to communicate with nodes inside the network. When a CN sends packets to a roaming user, it uses the MN's home IP address. As in conventional Mobile IP, the HA intercepts the packets and sends the encapsulated packet to the MN's current GW. The GW decapsulates the packet and forwards it to the MN's home address using a node-specific route. Thus, the nodes sending or receiving packets to and from the MN remain unaware of the node location inside the Cellular IP network.

HAWAII and Cellular IP are based on very similar concepts. There is a major difference between them, however. In the HAWAII protocol, most of the intelligence is in the network part, while in Cellular IP most of the intelligence is in the mobile node. Therefore, Cellular IP is not optimal for management of the security and quality of service. However, the network equipment is simpler and therefore cheaper.

4.5.3.3   RSVP Support for Mobile IPv6

The solution for the RSVP operation over IP tunnels is provided for Mobile IPv6 as well [159]. The specification in the draft, proposes three solutions:

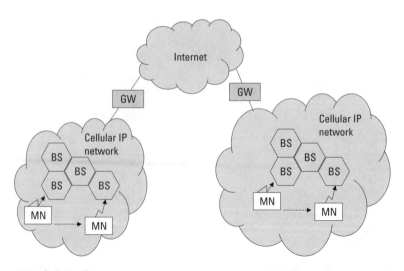

Figure 4.16 Cellular IP.

1. The first solution requires modifications in RSVP at both mobile and correspondent nodes, so that they have to be aware of Mobile IPv6 addressing.

2a. In order to enhance the performance and make handoffs smooth and seamless, optional *triggers/objects* are added to RSVP messages. The RSVP PATH messages are triggered on bindings updates and home address objects that are contained in RSVP RESV messages. Thus, intermediate routers are enabled to recognize connections and to use resources even when the COA changes.

2b. A *flow extension* mechanism is provided. This mechanism is able to extend the existing RSVP flows (i.e., flow_ids) that are applied on typical IP routers, to the new Mobile IP router. It is combined with a simultaneous binding option that has to be applied for the roaming mobile node. The mobile node receives packets on both previous and current COA.

To make Mobile IPv6 and RSVP interoperable, the minimal solution (1) is a requirement [159]. This requires the modification and the interfacing of the RSVP daemon and Mobile IP binding cache at both CN and MN.

The latter two solutions (2a or 2b) can provide uninterrupted operation since they support fast reestablishment or preservation of resource reservations when mobile nodes move. Table 4.3 presents a qualitative comparison of the latest two approaches.

It should be noted that triggers/objects is a quick solution with low complexity, which is able to provide sufficiently good service. The flow extension approach is a little more complex but has the advantage of faster deployment. In multiprovider environments, where the whole end-to-end path could not be controlled, a solution that modifies only CNs, access network routers, and MNs has a big advantage.

4.6 Relays for Next Generation 4G Systems

4.6.1 Introduction

Characteristics of future mobile/wireless radio networks will be quite different to those of current networks. Compared to a *uniform* spatial distribution of, say, voice traffic in a GSM cell, a much more nonuniform traffic will be witnessed in future networks in terms of spatial distribution within the cell of voice and data, in particular with even higher demands for a much broader bandwidth stretching from low to high bit rates for higher mobility. At the same time, the more we move to higher frequencies, the more the range of the cell shrinks compared

Table 4.3

RSVP Support for Mobile IPv6: Qualitative Comparison of Approaches 2a and 2b

Criteria	Triggers/Objects	Flow Extension
Changes to CN	Yes (required for minimal solution)	Yes (required for minimal solution)
Changes to Intermediate Routers	Yes (RSVP MIP object extension and reuse of flow's resources)	No
Changes to MIP-Router	No (forwarding of late packets is also an option here)	Yes (binding update interception, flow forwarding)
Changes to MN	Yes	Yes
Changes to HA	No	No
Supports Multicast Delivery	Yes	Yes
Bandwidth efficient	Yes	Yes (it is assumed efficient overprovisioning in the access network)
End-to-End Delay	Always shortest path (but reestablishment of resources requires a round-trip)	Slightly increased delay
Lossless HO	Yes (with forwarding of late packets)	Yes
HO Delay	Roundtrip	Faster
Implementation Complexity	Moderate	Higher

to current systems (e.g., GSM cells). This means that in order to cover the same geographical area, more BSs need to be deployed, which leads to much higher deployment costs for operators. Apart from those new needs, future network deployments will need to solve more adequately the problems that exist in current conventional network deployments and that are related to the issue of coverage; either extend/stretch the coverage of cells or provide better coverage in shadowed areas within a cell in terms of, say, better QoS. The above needs/requirements will need to be addressed either by new technologies (for instance, new air interface with conventional cellular architectures) or by novel concepts that will introduce new network elements and architectures.

One of those concepts that can address capacity and coverage issues is that of relaying. Relaying has attracted a lot of interest in recent years mainly through research projects of companies and universities, particularly the Wireless World Initiative New Radio (WINNER) project part of the EU IST FP6 [161] (see also Section 5.2.5.2). Relaying techniques can be classified according to several standpoints. *Analog relaying* refers to the simple case where the signal to be repeated is simply amplified and forwarded, possibly with a different frequency (frequency translation). The counterpart is *digital relaying*, where the signal is fully regenerated before being retransmitted (e.g., decode-and-forward). From a different perspective, *fixed* or *infrastructure relays* refer to repeating stations specifically put on fixed locations for that purpose. Relay stations can also be *infrastructureless*-that is, *movable* or *mobile*. The former refers to stations that can be moved to a specific location to help with some temporary requirements of coverage, or mounted on moving vehicles, while the latter refers to the use of other terminals as relaying stations. The relaying operation can be done in two hops, involving a single repeater, or in a truly multihop fashion, with several relaying stations being used. Fixed relaying techniques have been extensively studied (see [162, 163] and the references therein). Mobile relaying is currently being studied in the WINNER project and most of this section will deal with this subject. Mobile relays are effectively fixed-relay logical elements incorporating the mobility factor (see Figure 4.17). Although more complex in comparison to the fixed relaying approach, mobile relaying can offer incremental gains

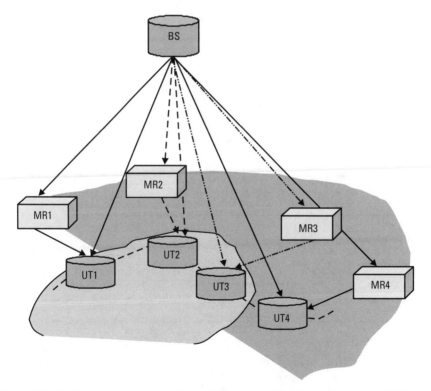

Figure 4.17 Mobile relay concept (MRi and UTi are the MR/UT positions for i=1,2,3,4 time instances).

in future networks, exactly due to that mobility. The main advantages are the plurality of types of mobile relays and the multiple locations where they can be found, which allow the network the opportunity to act on a more ad hoc basis to deal and address more nondeterministic needs.

4.6.2 Mobile Relay Types/Deployment Concepts

The first classification of mobile relays (MRs) (which can be done with reference to the ownership of the MR) is as follows:

- Dedicated mobile relays, with elements being built only for relaying purposes (Type I/II in Figure 4.18);
- User terminals (UT), which can act additionally as mobile relays (Type III in Figure 4.18).

Dedicated MRs are expected to be fitted on top of moving carriers (e.g., cars, ships, trains, or any other vehicle). As such, and with reference to the mobility of those MRs in relation to the mobility of the target area/UT population to cover, dedicated MRs can be split into two further categories:

- Dedicated MRs (Type I) to cover the UT population of the carrier on which they are fitted, either on the top or the inside (MR-UT mobility

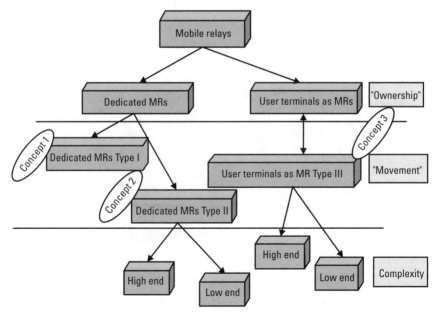

Figure 4.18 Types of mobile relays.

correlated; normally the same) (e.g., MR on a train to provide coverage inside the train);

- Dedicated MRs (Type II) to cover UT population outside the carrier they are fitted on (MR-UT mobility uncorrelated) (e.g., MR fitted on bus to provide coverage in a park).

Thus, in Figure 4.18 we present the main classification of MRs, as described in [164].

4.6.3 Rationale for Mobile Relays

Mobile relays in effect can address the same problems that fixed relays address: coverage and capacity. However, they are envisaged to cover cases that cannot or may not be adequately covered by fixed relays. In that sense fixed and mobile relay can be thought as complementing each other. Thus, the rationale for MR deployment can be summarized in the following points:

- To provide a cheaper solution (in terms of CAPEX/OPEX/backhaul costs) compared either to conventional topologies or fixed relay-based topologies.

- To provide coverage in unexpected events where a high concentration of UTs occurs (e.g., road accident/traffic jam). Those relays could be fitted on police cars or ambulances.

- To provide coverage for events that occur infrequently, but are known of in advance (e.g., a sporting event or concert). It might not be economically wise to deploy BSs next to a football stadium. In these cases MRs could be reused on demand.

- To provide coverage where fixed relays cannot go (e.g., provide coverage to terminals on top of ships). Thus, we rely on fixed relays, which, on such a carrier, become mobile relays.

- To reduce BS power transmission levels (by evenly distributing this power among mobile relays). Thus, we gain in increased capacity in, for example, a CDMA-based interference-limited system.

- To reduce the impact on the environment/humans. A plethora of GSM BSs and UMTS Node Bs has triggered a lot of controversy and complaints from communities and environmental organizations. Thus, through mobile relays we make deployment easier and cheaper (less cost is required to incorporate them on carriers rather than deploy BSs on top of buildings), and we provide less risk for humans in terms of power of RF emission, better aesthetical intervention, and less impact to the environment.

4.6.4 Applicability to Environments

Each of the aforementioned types of mobile relays is expected to have different characteristics and capabilities, especially in terms of mobility, complexity, and coverage. We expect them to be applicable or best suited to a number of different scenarios and environments [164].

- Type I: The main characteristics for Type I are high mobility, small coverage, and high complexity. Thus, they are envisaged to be typically applicable to scenarios of wide or rural areas.
- Type II: The main characteristics are medium/low mobility, medium/large coverage, and medium complexity. Thus, the applicability is towards hotspots and mostly wide area deployments.
- Type III: Typical characteristics are low/zero mobility, small coverage, and low/medium complexity. Thus, they are appropriate for indoors, hotspots, and possibly wide area deployments.

4.6.5 Parallels with Other Technologies

Technologies and concepts similar to the mobile relay concepts have been proposed in several fora, for existing and short/medium-term to-come systems. Extrapolating from these we can find commonalities in order to highlight the applicability of MR-based concepts on realistic scenarios.

- Positioning. A mobile relay in several locations can mimic a BS, which means that certain limitations of, say, using only round-trip time (RTT)/Cell Id positioning techniques in isolated sites are overcome and more techniques can be applicable [e.g., RTTs and observed time difference of arrival (OTDOA)]. At the same time, due to the favorable characteristics of the MRs (e.g., shorter range and better channel conditions), positioning accuracy is expected to be better compared to the same positioning techniques being applied on larger cells [165].
- In 3g Partnership Proejct (3GPP) the concept of repeaters has been proposed [166]. It will be interesting to see what type of simple mobile repeaters could be deployed and under what requirements.
- Vehicular networks are proposed as part of future networks for the automotive industry. Although more of an ad hoc/mesh-type networks compared to the MR-based hierarchical networks, interesting areas with regard

to interworking of those networks could be investigated. Effectively, MR-based networks could be seen as some kind of higher level layer (in the architecture concept) compared to the vehicular networks and effectively promote interworking between a mesh and a cellular network (e.g., mesh-network with gateway devices communicating with the MRs).

- Moving networks. For train companies/network operators the need for coverage inside trains is important. Thus, Type I MRs are very applicable for these types of deployments, and projects addressing this issue (the Moving Networks concept) are already underway [167].

- The concept of opportunity driven multiple access (ODMA), providing coverage through UTs to out-of-coverage UTs, was proposed in 3GPP [168]. It seems that now that networks can support more robust functionalities, faster signaling, ODMA could be extended to the Type III concept.

- Multimedia Broadcast/Multicast Services. MR-based networks could be simplified if MRs aimed to cover certain types of services. For instance, in the case of MBMS [169], the MRs do not need to maintain dedicated links with the terminals. As such, the whole deployment is much easier to implement. Furthermore, by allocating MBMS for the edge of cell on MRs, we free capacity in the BS.

In order to be able to propose specific solutions, a number of issues need to be resolved like the multiple access (MA) technique (e.g., OFDMA based), the duplex scheme (e.g., TDD), the level of complexity of a mobile relay (layer 1 or layer 2 mobile relays), the number of hops (preferably 2), routing and forwarding issues, security, signaling issues, types of protocols (e.g., nonstatic protocols, adaptive or feedback based), power control, and connectivity issues. Some of these have already been addressed, but further studies are required [167, 170]. However, in future networks it is envisaged that more complex functionalities and processes could be supported (e.g., faster signaling, higher processing power, shorter delays), which means that mobile relay-based concepts would be more easily supported compared to the point of view we have by looking at their applicability on current networks.

4.6.6 Cooperative Mobile Relaying

Cooperative relaying (CR) is another concept within the relaying field. The main idea is to use multiple relays to enhance the received signal in the UT. CR schemes could be seen under the mobile relaying concept, and thus, they give rise to the cooperative mobile relaying (CMR) concept [167]. Again, due to the different types of mobile relays, different levels of applicability exist for CMR schemes for a number of scenarios. Specifically, and with reference to the criteria of coverage, velocity, and types of applications, we prioritize the three types of MR with reference to the CMR concept.

- For short-type/fast applications/services, supporting two-hop strategy, Type III MRs are promising in providing CMR. The main advantage is the plurality of terminals in multiple locations out of which we can select the optimum to provide cooperative schemes. The main restrictions seem to be the frequent positioning, the low computational power, the possible reduction of the user experience (e.g., battery life), and the small coverage, which is, however, compensated by the low/zero mobility.

- For Type II, the problem is the relatively high mobility (average 40 km/h) but this can be compensated by the large coverage of the MR (up to 500m), which means that connectivity will be in general similar to that of Type III. So, these types of MRs could be applicable for more bandwidth-hungry applications/services due to the higher complexity incorporated into them.

- CMR may not be that applicable for Type I, due to the very good links we assume between MR-BS and MR-UTs, which is the initial reason for applying CMR (i.e., poor links). So, even though CMR could be supported it is not envisaged to be of high incremental gain to the performance. However, under a possible MIMO approach (multiply the bit rates by x times) it could be applicable.

4.6.7   Conclusions

Mobile relays have been proposed as part of the relay-based topologies for future network deployments for 4G systems in order to provide better coverage and increased capacity in cellular/wireless environments. A number of potential gains can be extracted by the use of mobile relays, as a complementary solution (e.g., on an ad hoc basis) to that of fixed relays. As we have seen, mobile relays pose a number of challenges, mainly those related to the issue of mobility and extending to routing, forwarding, positioning, connectivity, security, and complexity. At the same time, though, as we have analyzed, mobile relays could potentially better cover certain needs compared to conventional or fixed relay-based topologies by addressing and incorporating other concepts (e.g., reconfigurability, to be able to cope in several locations for multiple RATs).

4.7 Other Enabling Technologies

Other technologies have been identified as fundamental for allowing the development of future 4G. We now list some other important techniques and technologies not considered in this chapter: adaptive modulation and coding, techniques for seamless vertical and horizontal handovers, cross-layer design and optimization, multidimensional scheduling (time, frequency, space), techniques for reducing the PAPR problem typical of multicarrier systems, sensor networks, battery technology, and access techniques specifically tailored for short-range communications [171]. The latter techniques have received considerable attention lately. These are basically ultra-wideband (UWB) and optical wireless communication techniques. Interested readers are referred to [172, 173] and the references therein for a comprehensive introduction to UWB. The role of optical wireless systems within future 4G communications is discussed in detail in [174]. Optical wireless communications is in fact a very attractive complementary technology that has been extensively considered recently due to its unmatched advantages [175].

References

[1]   Alamouti, S., "Simple Transmit Diversity Technique for Wireless Communications," IEEE JSAC, Vol. 16, No. 8, October 1998, pp. 1451-1458.

[2]   Foschini, G., "Layered Space-Time Architecture for Wireless Communication in a Fading Environment when Using Multi-Element Antennas," Bell Labs. Tech. L, 1996, pp. 41-59.

[3]   Tarokh, V., N. Seshadri, and A. Calderbank, "Space-Time Codes for High Data Rate Wireless Communications: Performance Criterion and Code Construction," IEEE Trans. Information Theory, Vol. 44, March 1998, pp. 744-765.

[4]   Jafarkhani, H., "A Quasiorthogonal Space-Time Block Code," IEEE Trans. Communications, Vol. 49, 2001, pp. 1-4.

[5]   Zheng, L., and D.N.C Tse, "Diversity and Multiplexing: A Fundamental Trade-Off in Multiple-Antenna Channels," IEEE Trans. Information Theory, Vol. 49, May 2003, pp. 1073-1096.

[6]   Heath, Jr., R. W., and A. Paulraj, "Switching Between Spatial Multiplexing and Transmit Diversity Based on Constellation Distance," Proc. of Allerton Conf. on Communication Cont. and Comp., October 2000.

[7]   Skjevling, H., D. Gesbert, and N. Christophersen, "Combining Space Time Block Codes and Multiplexing in Correlated MIMO Channels: An Antenna Assignment Strategy," Proc. of Nordic Signal Processing Conference (NORSIG), June 2003.

[8]   Gorokhov, A., D. Gore, and A. Paulraj, "Diversity Versus Multiplexing in MIMO System with Antenna Selection," Allerton Conference, October 2003.

[9]   Xin, Y., Z. Wang, and G. B. Giannakis, "Space-Time Diversity Systems Based on Linear Constellation Precoding," IEEE Trans. Com., Vol. 49, 2001, pp. 1-4.

[10]   Ma, X., and G. B. Giannakis, "Complex Field Coded MIMO Systems: Performance, Rate, and Trade-Offs," Wirel. Commun. Mob. Comput., 2002, pp. 693-717.

[11]   Jung, T. J., and K. Cheun, "Design of Concatenated Space-Time Block Codes Using Signal Space Diversity and the Alamouti Scheme," IEEE Com. Letters, Vol. 7, July 2003, pp. 329-331.

[12]   Chung, S., T. J. Richardson, and R. Urbanke, "Analysis of Sum-Product Decoding of Low-Density Parity-Check Codes Using a Gaussian Approximation," IEEE Trans on Information Theory, Vol. 47, No. 2, February 2001, pp. 657-670.

[13]   Zafar, M., A. Khan, and B. S. Rajan, "Space-Time Block Codes from Coordinate Interleaved Orthogonal Designs," IEEE ISIT, 2002, p. 275.

[14]   Berrou, C., A. Glavieux, and P. Thitimajshima, "Near Shannon-Limit Error-Correction Coding and Decoding: Turbo-Codes," Proc. Int. Communication Conf., Geneva, Switzerland, May 1993, pp. 1064-1070.

[15]   Hagenauer, J., E. Offer, and L. Papke, "Iterative Decoding of Binary Block and Convolutional Codes," IEEE Trans. on Information Theory, Vol. IT-42, No. 2, March 1996, pp. 429-445.

[16]   Sipser, M., and D. A. Spielman, "Expander Codes," IEEE Trans. Inform. Theory, Vol. 42, November 1996, pp. 1710-1722.

[17]   MacKay, D. J. C., and R. M. Neal, "Near Shannon Limit Performance of Low-Density Parity-Check Codes," Electron. Lett., Vol. 32, August 1996, pp. 1645-1646.

[18]   MacKay, D. J. C., "Good Error-Correcting Codes Based on very Sparse Matrices," IEEE Trans. Inform. Theory, Vol. 45, March 1999, pp. 399-431.

[19]   Gallager, R. G., Low Density Parity-Check Codes, Cambridge, MA: MIT Press, 1963.

[20]   Benedetto, S., and G. Montorsi, "Unveiling Turbo Codes: Some Results on Parallel Concatenated Coding Scheme," IEEE Trans. on Information Theory, Vol. 42, March 1996, pp. 409-428.

[21]   Duman, T. M., and M. Salehi, "New Performance Bounds for Turbo Codes," IEEE Trans. on Communications, Vol. 46, June 1998, pp. 717-723.

[22]   Sason, I., and S. Shamai, "On Improved Bounds on the Decoding Error Probability of Block Codes over Interleaved Fading Channel, with Application to Turbo-Like Codes," IEEE Trans. on Information Theory, Vol. 47, September 2001, pp. 2275-2299.

[23]   Pearl, J., Probabilistic Reasoning in Intelligent Systems: Networks of Plausible Inference, Morgan Kaufmann, 1988.

[24]   McEliece, R. J., D. J. C. MacKay, and J.-F. Cheng, "Turbo Decoding as an Instance of Pearl's Belief Propagation Algorithm," IEEE Journal on Selected Areas in Communications, Vol. 16, No. 2, February 1998, pp. 140-152.

[25]   Luby, M. G., et al., "Analysis of Low Density Codes and Improved Designs Using Irregular Graphs," Proc. of Annual ACM Symposium on Theory of Computing, 1998, pp. 249-258.

[26] Richardson, T., and R. Urbanke, "Efficient Encoding of Low-Density Parity-Check Codes," *IEEE Trans. on Information Theory*, Vol. 47, February 2000, pp. 638-656.

[27] Richardson, T. J., and R. Urbanke, "The Capacity of Low-Density Parity-Check Codes Under Message-Passing Decoding," *IEEE Trans. on Information Theory*, Vol. 47, No. 2, February 2001, pp. 599-618.

[28] Richardson, T. J., and R. Urbanke, "Design of Capacity Approaching Irregular Low-Density Parity Check Codes," *IEEE Trans. on Information Theory*, Vol. 47, No. 2, February 2001, pp. 619-637.

[29] Bahl, L. R., et al., "Optimal Decoding of Linear Codes for Minimizing Symbol Error Rate," *IEEE Trans. on Information Theory*, March 1974, pp. 284-287.

[30] Goldsmith, A., and P. Varaiya, "Increasing Spectral Efficiency Through Power Control," *Proc. of ICC'93*, June 1993, pp. 600-604.

[31] Cover, T. M., *Elements in Information Theory*, New York: Wiley Interscience, 1991.

[32] Goldsmith, J., and S. Chua, "Adaptive Coded Modulation for Fading Channels," *IEEE Trans. on Communications*, Vol. 46, No. 5, May 1998, pp. 595-602.

[33] Wicker, S. B., *Error Control Systems for Digital Communication and Storage*, Englewood Cliffs, NJ: Prentice Hall, 1995.

[34] Chase, D., "Code Combining - A Maximum-Likelihood Decoding Approach for Combining an Arbitrary Number of Noisy Packets," *IEEE Trans. on Communication*, Vol. 33, No. 5, 1985, pp. 385-393.

[35] Harvey, B. A., and S. B. Wicker, "Packet Combining Systems Based on the Viterbi Decoder," *IEEE Trans. on Communications*, Vol. 42, No. 2-4, May 1995, pp. 1544-1557.

[36] Souissi, S., and S. B. Wicker, "A Diversity Combining DS/CDMA System with Convolutional Encoding and Viterbi Decoding," *IEEE Trans. on Vehicular Technology*, Vol. 44, No. 2, May 1995, pp. 304-312.

[37] Hagenauer, J., "Rate-Compatible Punctured Convolutional Codes (RCPC Codes) and Their Applications," *IEEE Trans. on Communication* Vol. 36, December 1988, pp. 389-400.

[38] Lugand, L. R., D. J. Costello, and R. H. Deng, "Parity Retransmission Hybrid ARQ Using Rate 1/2 Convolutional Codes on a Nonstationary Channel," *IEEE Trans. on Communication*, Vol. COM-37, No. 7, July 1989, pp. 755-765.

[39] Kallel, S., and D. Haccoun, "Generalized Type II Hybrid ARQ Scheme Using Punctured Convolutional Coding," *IEEE Trans. on Communication*, Vol. 38, No. 11, November 1990, pp. 1938-1946.

[40] Chae, C. B., et al., "A Constellation-Rotation Based Precoding Approach Enabling Reduced Complexity Full-Diversity Full-Rate Space Time Block Codes," WWRF12, Toronto, November 2004.

[41] Lutkepohl, H., Handbook of Matrices, New York: John Wiley & Sons, 1996.

[42]   Tarokh, V., H. Jafarkhani, and A. R. Calderbank "Space-Time Block Coding for Wireless Communications: Performance Results," IEEE J. Selec. Areas Commun., Vol. 17, March 1999, pp. 451-460.

[43]   Ganesan, G., and P. Stoica, "Space-Time Block Codes: A Maximum SNR Approach," IEEE Tr. Inf. Theory, Vol. 47, May 2001, pp. 1650-1656.

[44]   Brink, S., "Convergence of Iterative Decoding," IEE Electronics Letters, Vol. 35, No. 10, May 1999, pp. 806-807.

[45]   Brink, S., "Convergence Behavior of Iteratively Decoded Parallel Concatenated Codes," IEEE Trans. on Communication, Vol. 49, No. 10, October 2001, pp. 1727-1737.

[46]   Gameld, H. E., and A. R. Hammons, Jr., "Analyzing the Turbo Decoder Using the Gaussian Approximation," IEEE Trans. on Information Theory, Vol. 47, No. 2, February 2001, pp. 671-686.

[47]   Divsalar, D., S. Dolinar, and F. Pollara, "Iterative Turbo Decoder Analysis Based on Density Evolution," IEEE Journal on Selected Areas in Communication, Vol. 19, No. 5, May 2001, pp. 891-907.

[48]   Richardson, T. J., and R. Urbanke, "Threshold for Turbo Codes," Proc. of ISIT'2002, Sorrento, Italy, June 2000, p. 317.

[49]   Jorguseski, L., J. Farserotu, and R. Prasad, "Radio Resource Allocation in 3rd Generation Mobile Communication Systems," IEEE Communications Magazine, Vol. 39, No. 2, February 2001, pp. 117–123.

[50]   Zander, J., "Radio Resource Management - An Overview," IEEE VTC '96, Vol. 1, May 1996, pp. 16-20.

[51]   Zander, J., "Radio Resource Management in Future Wireless Networks: Requirements and Limitations," IEEE Communication Magazine, August 1997, pp. 30-36.

[52]   Zander, J., "Radio Resource Management in 3rd Generation Personal Communication Systems," IEEE Communication Magazine , No. 8, August 1998.

[53]   3GPP Technical Specification Group, "QoS Concept and Architecture," (3GTS23.107), October 1999.

[54]   3GPP Specification TS 23.107, "UMTS QoS Concept and Architecture," R'99/R4.

[55]   GSM 02.60, "GPRS Service Description - Stage 1," Version 7.5.0, Release 1998.

[56]   Stuckmann, P., and O. Paul, "Dimensioning GSM/GRPS Networks for Circuit-Switched and Packet-Switched Services," Proc. WPMC '01, Aalborg, Denmark, September 2001, pp. 597–602.

[57]   Holma, H., and A. Toskala, WCDMA for UMTS-Radio Access for Third Generation Mobile Communications, New York: John Wiley & Sons, 2001.

[58]   3GPP TS 25.211, "Physical Channels and Mapping of Transport Channels onto Physical Channels (FDD)," v.5.1.0, June 2002.

[59]   3GPP TR 25.848, "Physical Layer Aspects of UTRA High Speed Downlink Packet Access," v4.0.0, March 2001.

[60]   Parkvall, S., et al., "Evolving WCDMA for Improved High Speed Mobile Internet," *Future Telecommunication Conference*, Beijing, China, November 28-30, 2001.

[61]   ETSI TS 101 475, "Broadband Radio Access Network (BRAN); HIPERLAN Type 2; Physical (PHY) layer," v1.1.1, April 2000.

[62]   ETSI TS 101 761-1, "Broadband Radio Access Network (BRAN); HIPERLAN Type 2; Data Link Control (DLC) Layer; Part1: Basic Data Transport Functions," v1.1.1, April 2000.

[63]   Doufexi, A., et al., "A Comparison of the HIPERLAN/2 and IEEE 802.11a Wireless LAN Standards," *IEEE Communications Magazine*, May 2002, Vol. 40, Issue 5, May 2002, pp. 172–180.

[64]   ETSI TS 101 761-2, "Broadband Radio Access Network (BRAN); HIPERLAN Type 2; Data Link Control (DLC) Layer; Part 2: Radio Link Control (RLC) Sublayer," v1.1.1, April 2000.

[65]   IEEE Std. 802.11 a/D7.0-1999, "Part 11: Wireless LAN Medium Access Control (MAC) and Physical Layer (PHY) Specifications: High Speed Physical Layer in the GHz Band."

[66]   Gupta, P., and P. R. Kumar, "The Capacity of Wireless Networks, " *IEEE Transactions on Information Theory*, Vol. 46, No. 2, March 2000, pp. 388-404.

[67]   Zahedi, A., and K. Pahlavan, "Capacity of a Wireless LAN with Voice and Data Services," *IEEE Transactions on Communications*, Vol. 48, No. 7, July 2000, pp.34–40.

[68]   Sinha, P., R. Sivakumar, and V. Bharghavan, "CEDAR: A Core-Extraction Distributed ad hoc Routing algorithm," *IEEE Infocom '99*, New York, March 1999, pp. 202–209.

[69]   Lin, C. R., and M. Gerla, "MACA/PR: An Asynchronous Multimedia Multi-Hop, Wireless Network," *Proceedings of IEEE INFOCOM '97*, 1997.

[70]   Berg, M. "A Concept for Hybrid Random/Dynamic Radio Resource Management," *Proc. IEEE PIMRC'98*, Boston, 1998, pp. 424–428.

[71]   Mihailescu, C., X. Lagrange, and P. Godlewski, "Dynamic Resource Allocation in Locally Centralized Cellular Systems," Proc. VTC'98, Ottawa, Canada, 1998, pp. 1695-1700.

[72]   Katezela, I., and M. Naghshineh, "Channel Assignment Schemes for Cellular Mobile Telecommunication Systems: Comprehensive Survey," IEEE Personal Communications, Vol. 3, No. 3, June 1996, pp.10–31.

[73]   Zander, J., "Performance Bounds for Joint Power Control & Link Adaptation for NRT Bearers in Centralized (Bunched) Wireless Networks," Proc. PIMRC'99, September 1999.

[74]   Zander, J., "Trends in Resource Management Future Wireless Networks," *IEEE Wireless Communications and Networking Conference*, WCNC. 2000, Vol. 1, pp. 159-163.

[75]   Zander, J., S.-L. Kim, and M. Almgren, *Radio Resource Management for Wireless Networks*, Norwood, MA: Artech House, 2001.

[76]   Mitila, J., "The Software Radio Architecture," IEEE Commun. Mag., May 1995, pp. 26-38.

[77]   Lackey, R. J., and D. W. Vpmal, "Speakeasy: The Military Software Radio," IEEE Communication Magazine, May 1995, pp. 56-61.

[78] European Commission DG XIII-B, Proc. of Software Radio Workshop, May 1997.

[79] Special Issue on Globalization of Software Radio, IEEE Communication Magazine February 1999.

[80] Special Issue on Software Radio, IEEE Personal Communications, August 1999.

[81] Special Issue on Software and DSP in Radio, IEEE Communication Magazine, August 2000.

[82] Special Issue on Software Defined Radio and Its Technologies, IEICE Trans. on Communication June 2000.

[83] Kohno, R., "Structures and Theories of Software Antennas for Software Defined Radio," IEICE Trans. Communication, Vol. E83-B, No. 6, June 2000, pp. 1189-1199.

[84] Karasawa, Y., "Algorithm Diversity in a Software Antenna," IEICE Trans. Communication Vol. E83-B, No. 6, June 2000, pp. 1229-1236.

[85] Magngum, C., "Market Opportunities II: The MMITS Forum Market Forecast Study," First International Software Radio Workshop, Rohdes, Greece, June 1998, pp. 15-24.

[86] Harada, H., and M. Fujise, "Multimode Software Radio System by Parameter Controlled and Telecommunication Toolbox Embedded Digital Signal Processing Chipset," Proc. of 1998 ACTS Mobile Communications Summit, June 1998, pp. 115-120.

[87] Harada, H., Y. Kamio, and M. Fujise, "Multimode Software Radio System by Parameter Controlled and Telecommunication Component Block Embedded Digital Signal Processing Hardware," IEICE Trans. Communication Vol. E83-B, No. 6, June 2000.

[88] Harada, H., "A Proposal of Multi-Mode & Multi-Service Software Radio Communication Systems for Future Intelligent Telecommunication Systems," Proc. of International Symposium on Wireless Personal Multimedia Communications (WPMC'99), September 1999, pp. 301-304.

[89] Harada, H., "Wireless Terrorism," Denpa-Shinbun, May 30, 2000.

[90] Tsurumi, H., and Y. Suzuki, "Broadband RF Stage Architecture for Software-Defined Radio in Handheld Terminal Applications," IEEE Commun. Mag., Vol. 37, No. 2, February 1999, pp. 90-95.

[91] Perkins, C. E., (Ed.), "IP Mobility Support," RFC2002, proposed standard, IETF Mobile IP Working Group, October 1996.

[92] Schooler, E., et al., "SIP: Session Initiation Protocol," IETF RFC 2543, March 1999.

[93] Perkins, C., E., "Mobile IP," IEEE Communications Magazine, May 1997, Vol. 35, Issue 5, pp. 84–99.

[94] Deering, S. E., (Ed.), "ICMP Router Discovery Messages," RFC 1256, September 1991.

[95] Perkins, C., "Minimal Encapsulation Within IP," RFC2004, October 1996.

[96] Hanks, S., et al., "Generic Routing Encapsulation over IPv4 Networks," RFC 1701, October 1994.

[97] Jacobson, V., "Compressing TCP/IP Headers for Low-Speed Serial Links," RFC 1144, February 1990.

[98] Rivest, R., "The MD5 Message-Digest Algorithm," RFC 1321, April 1992.

[99] Perkins, C., "IP Encapsulation Within IP," RFC2003, October 1996.

[100] Plummer, D. C., "An Ethernet Address Resolution Protocol: Or Converting Network Protocol Addresses to 48.bit Ethernet Addresses for Transmission on Ethernet Hardware," RFC 826, November 1982.

[101] Huitema, C., *Routing in the Internet*, Englewood Cliffs, NJ: Prentice Hall, 1995.

[102] Perkins, C., and P. Bhagwat, "Highly Dynamic Destination-Sequenced Distance Vector Routing (DSDV) for Mobile Computers," Proc. ACM SIFCOMM, 1996.

[103] Johnson, D. B., and C. Perkins, "Dynamic Source Routing in Ad Hoc Wireless Networks," Mobile Computing, Kluwer Academic Publishers, 1996.

[104] Perkins, C., and B. J. Johnson, "Route Optimization in Mobile IP," Internet Draft, http://draft-ietf-mobileip-optim-10.txt, work in progress, November 2000.

[105] Perkins, C., E., "Mobile Networking Through Mobile IP," IEEE Internet Computing, 1998, Vol. 2, Issue 1, pp. 58–69.

[106] Braden, R., D., Clark, and S. Shenker, "Integrated Services in the Internet Architecture: An Overview," RFC1633, June 1994.

[107] Braden, R., et al., "Resource Reservation Protocol (RSVP) - Version 12 Functional specification," available at http://www.ietf.org/html.charters/intserv-charter.htm, August 12, 1996.

[108] Ors, T., and C. Rosenberg, "Providing IP QoS over GEO Satellite Systems Using MPLS," International Journal of Satellite Communications, Vol. 19, 2001, pp. 443-461.

[109] Finenberg, V., "A Practical Architecture for Implementing End-to-End QoS in an IP Network," IEEE Communications Magazine, January 2002, pp. 122-130.

[110] Ferguson, P., and G. Huston, *Quality of Service-Delivering QoS on the Internet and in Corporate Networks*, New York: John Wiley & Sons, 1998.

[111] ITU Draft Recommendation I.ipatm, "IP over ATM," September 1999.

[112] Le Faucher, F., "MPLS Support of Differentiated Services," Internet-draft, IETF MPLS Working Group, March 2000.

[113] Jamoussi, B., (Ed.), "Constraint-Based LSP Setup Using LDP," Internet-draft, IETF MPLS Working Group, September 1999.

[114] Terzis, A., et al., "RSVP Operation over IP Tunnels," RFC2746, January 2000.

[115] Andreoli, G., et al., "Mobility Management in IP Networks Providing Real-Time Services," Proc. Annual International Conference on Universal Personal Communications, 1996, pp. 774-777.

[116] Mahadevan, I., and M. Sivalingham, "An Architecture for QoS Guarantees and Routing in Wireless/Mobile Networks," ACM Intl. Workshop on Wireless and Mobile Multimedia, 1998.

[117] Chen, W.-T., and L.-C. Huang, "RSVP Mobility Support: A Signaling Protocol for Integrated Services Internet with Mobile Hosts," INFOCOM 2000, Vol. 3, 2000, pp. 1283-1292.

[118] Alam, M., R. Prasad, and J. R. Farserotu, "Quality of Service Among IP-Based Heterogeneous Networks," *IEEE Personal Communications,* December 2001, Vol. 18, Issue 6, pp. 18–24.

[119] Bernet, Y., "The Complementary Roles of RSVP and Differentiated Services in the Full-Service QoS Network," IEEE Communication Magazine, February 2000.

[120] Rouhana, N., and E. Horlait, "Differentiated Services and Integrated Services Use of MPLS," Proc. IEEE ISCC2000, pp. 194-199.

[121] Ford, W., Computer Communications Security-Principles, Standard Protocols and Techniques, Englewood Cliffs, NJ: Prentice Hall, 1994.

[122] Kaufman, C., R. Perlman, and M. Speciner, Network Security - Private Communication in a Public World, Englewood Cliffs, NJ: Prentice Hall, 1995.

[123] Stalling, W., *Cryptography and Network Security: Principle and Practice,* Second Edition, Englewood Cliffs, NJ: Prentice Hall, 1998.

[124] Jefferies, N., C. Mitchell, and M. Walker, "A Proposed Architecture for Trusted Third Party Services," in *Cryptography: Policy and Algorithms,* E. Dawson and J. Golic, Springer-Verlag LNCS 1029, 1996, pp. 98-104.

[125] Kent, S., and R. Atkinson, "IP Authentication Header," RFC 2402, November 1998.

[126] Thayer, R., N. Doraswamy, and R. Glenn, "IP Security Document Roadmap," RFC 2411, November 1998.

[127] Rigney, C., et al., "Remote Authentication Dial In User Service (RADIUS)," RFC 2138, April 1997.

[128] Calhoun, R., "DIAMETER," Internet-draft, http://draft-calhoun-diameter-07.txt, work in progress, November 1998.

[129] Perkins, C., "Mobile IP Joins Forces with AAA," IEEE Personal Communications, August 2000.

[130] Montenegro, G., "Reverse Tunnelling for Mobile IP," RFC 3024, January 2001.

[131] Teo, W. T., Y. Li, "Mobile IP Extension for Private Internet Support (MPN)," Internet-draft, http://draft-teoyli-mobileip-mvpn-02.txt, work in progress, 1999.

[132] Kent, S., and R. Atkinson, "Security Architecture for the Internet Protocol," Internet-draft, http://draft-ietf-ipsec-arch-sec-02.txt, work in progress, November 1997.

[133] Zao, J. K., and M. Condell, "Use of IPSec in Mobile IP," Internet-draft, http://draft-ietf-mobileip-ipsec-use-00.txt, work in progress, November 1997.

[134] Perkins, C., D. B. Johnson, "Registration Keys for Route Optimization," Internet-draft, http://draft-ietf-mobileip-regkey-03.txt, work in progress, July 2000.

[135] Diffie, W., and M. Hellman, "New Directions in Cryptography," IEEE Transactions on Information Theory, Vol. 22, November 1976, pp. 644-654.

[136] Perkins, C. E., and P. R. Calhoun, "Mobile IP Challenge/Response Extensions," RFC 3012, November 2000.

[137] Jacobs, S., "Mobile IP Key Based Authentication," Internet-draft, http://draft-jacobs-mobileip-pki-auth-02.txt, work in progress, March 1999.

[138] Sufatrio, K. Y. L., "Mobile IP Registration Protocol: A Security Attack and New Secure Minimal Public-key Based Authentication," SPAN '99, June 1999, pp. 364-369.

[139] Calhoun, P. R., and C. E. Perkins, "Mobile IP Network Address Identifier Extension," Internet-draft, http://draft-ietf-mobileip-mn-nai-01.txt, work in progress, May 1999.

[140] Aravamudhan, L., M. R. O'Brien, and B. Patil, "NAI Resolution for Wireless Networks," Internet-draft, http://draft-ietf-mobileip-nai-wn-00.txt, work in progress, February 1999.

[141] Calhoun, P. R., and A. C. Rubens, "DIAMETER Reliable Transport Extensions," Internet-draft, http://draft-calhoun-diameter-mobileip-01.txt, work in progress, February 1999.

[142] Montenegro, G., and Gupta, V., "Firewall support for Mobile IP," Internet-draft, http://draft-montenegro-firewall-sup-03.txt, work in progress, January 1998.

[143] Leech, M., et al., "SOCKS Protocol Version 5," RFC 1928, March 1926.

[144] Aziz, A., and M. Patterson, "Design and Implementation of SKIP," available at http://skip.incog.com/inet-95.ps, 1995.

[145] Mink, S., et al., "FATIMA: A Firewall-Aware Transparent Internet Mobility Architecture," ISCC'2000, July 2000, pp. 172-179.

[146] Johnson, D. B., and C. Perkins, "Mobility Support in IPv6," Internet-draft, http://draft-ietf-mobileip-ipv6-13.txt, work in progress, November 2000.

[147] Ferguson, P., and D. Senie, "Network Ingress Filtering: Defeating Denial of Service Attacks which Employ IP Source Address Spoofing," RFC 2267, January 1998.

[148] Narten, T., E. Nordmark, and W. A. Simpson, "Neighbour Discovery for IP version 6 (IPv6)," RFC 1970, August 1996.

[149] Thomson, S., and T. Narten, "IPv6 Stateless Address Autoconfiguration," RFC 1971, August 1996.

[150] Caceres, R., and V. N. Padmanabhan, "Fast and Scalable Handoffs for Wireless Internetworks," Proc. MOBICOM '96, ACM, August 1996, pp. 76-82.

[151] Perkins, C., "Mobile IP Local Registration with Hierarchical Foreign Agents," IETF Internet-draft, work in progress, February 1996.

[152] Das, S., A. Misra, and P. Agrawal, "TeleMIP: Telecommunications-Enhanced Mobile IP Architecture for Fast Intradomain Mobility," IEEE Personal Communications, Vol. 7 August 2000, pp. 50-58.

[153] Hiller, T., (Ed.), "Wireless IP Network Architecture Based on IETF Protocols," Ballot v. PN-4286, TIA/TR45, June 1999.

[154] Perkins, C., and P. R. Calhoun, "Mobile IP Network Access Identifier Extension for IPv4," Internet-draft, http://draft-ietf-mobileip-mn-nai-07.txt, work in progress, July 1999.

[155] Geiger, R. L., J. D. Solomon, and K. J. Crisler, "Wireless Network Extension Using Mobile IP," IEEE Micro, Vol. 17, No. 6, 1997, pp. 63-68.

[156] La Porta, T., R. Ramjee, and L. Li, "IP Micro-Mobility Support Using HAWAII," Internet-draft, http://draft-ietf-mobileip-hawaii-00.txt, work in progress, June 1999.

[157] Valko, A. G., "Cellular IP: A New Approach to Internet Host Mobility," Comp. Commun. Rev., January 1999, pp. 50-65.

[158] Wan, C.-Y., et al., "Cellular IP," Internet-draft, http://draft-valko-cellularip-01.txt, work in progress, October 1999.

[159] Fankhauser, G., S. Hadjiefthymiades, and N. Nikaein, "RSVP Support for Mobile IP Version 6 in Wireless Environments," Internet-draft, http://draft-fhns-rsvp-support-in-mipv6-00.txt, November 1998.

[160] Gosh, D., V. Sarangan, and R. Acharya, "Quality-of-Service Routing in IP Networks," IEEE Transactions on Multimedia, Vol. 3, No. 2, June 2001, pp. 200-208.

[161] https://www.ist-winner.org/.

[162] Yanikomeroglu, H., "Fixed and Mobile Relaying Technologies for Cellular Networks," Second Workshop on Applications and Services in Wireless Networks (ASWN'02), Paris, France, July 3-5, 2002, pp. 75-81.

[163] Pabst, R., et al., "Relay-Based Deployment Concepts for Wireless and Mobile Broadband Radio," IEEE Communications Magazine, Vol. 42, No. 9, September 2004, pp. 80-89.

[164] IST WINNER D3.4, "Description and Assessment of Relay Based Cellular Deployment Concepts for Future Radio Scenarios Considering 1st Protocol Characteristics," June 2005.

[165] 3GPP TS 25.305, "Stage 2 Functional Specification of User Equipment (UE) Positioning in UTRAN."

[166] 3GPP TS 25.956, "UTRA Repeater Planning Guidelines and System Analysis."

[167] IST WINNER D3.2, "Description of Identified New Relay Based Radio Network Deployment Concepts and First Assessment by Comparison Against Benchmarks of Well Known Deployment Concepts Using Enhanced Radio Interface Technologies," February 2005.

[168] 3GPP TR 25.924, "Opportunity Driven Multiple Access."

[169] 3GPP TS 22.146, "Multimedia Broadcast/Multicast Services."

[170] Bakaimis, B., and T. Lestable, "Connectivity Investigation of Mobile Relays for Next Generation Wireless Systems," VTC'05 Spring, Stockholm, Sweden, June 2005.

[171] Katz, M., and F. Fitzek, "On the Definition of the Fourth Generation Wireless Communications Networks: The Challenges Ahead," Proceedings of International Workshop on Convergent Technologies (IWCT'2005), Center for Wireless Communications, University of Oulu, Finland, June 6-10, 2005.

[172] Porcino, D., and W. Hirt, "Ultra-Wideband Radio Technology: Potential and Challenges Ahead," IEEE Communications Magazine, Vol. 41, No. 7, July 2003, pp. 66-74.

[173] Allen, B., "White Paper on UWB for Short-Range Communications," WWRF, Working Group 5 on short-range communications.

[174] O'Brien, D., and M. Katz, "White Paper: Short-Range Optical Wireless Communications," WWRF, Working Group 5 on short-range communications, Oct. 2004.

[175] O'Brien, D., and M. Katz, "Optical Wireless Communications within Fourth-Generation Wireless Systems," Journal of Optical Networking, Vol. 4, No. 6, June 2005, pp. 313-322.

5

4G Research Initiatives and Developments

5.1 Introduction

Defining a sound technical framework for future 4G wireless systems and determining the appropriate enabling technologies is undoubtedly a colossal task, requiring open discussions and collaboration among a large number of parties. Key players in this multiyear development process are manufacturers, academia, wireless operators, service providers, regulatory bodies, and governmental agencies, working on their own and in collaborative projects. Currently, many activities are being conducted that aim not only at defining future 4G systems but also at determining the most promising technical solutions for implementing such systems. In this chapter research initiatives and development work focused on 4G systems are introduced and briefly discussed. First, the scope and goals of the most relevant large-scale research initiatives dealing with 4G will be briefly described. Then, the development work currently being undertaken mainly by industry to develop concrete 4G solutions is described. 4G development work is highly fragmented, and, in general, the work scope of each organization varies greatly, depending on its core business and size. Current efforts can be regarded as being part of a preceding stage to the upcoming 4G standardization process, hence the importance of these activities.

5.2   Major Research Initiatives Focusing on 4G

In this section the most prominent research initiatives on 4G systems are presented. Particularly attention is given to wide-scope multipartner projects aiming to achieve a global footprint. Based on the number and size of ongoing projects, and taking into account the number of participating partners, Asia and Europe appear to be more involved in 4G-related research than any other area. In addition to the main initiatives, a list of other important projects is included at the end of this section.

5.2.1   Wireless World Research Forum

Established in 2001 by a number of leading telecom manufacturers, the Wireless World Research Forum (WWRF) [2] is a nonprofit organization that aims to create a common understanding of visions and key technologies related to future wireless networks. Despite its European origins, the WWRF has succeeded in extending its range of activities and participation into Asia and America. With key participating organizations and a membership count exceeding 150, it can be stated with certainty that the WWRF is now the largest research initiative focusing on 4G systems. In addition to heavyweight telecom manufactures, numerous IT companies, and a large number of universities from across the globe, network operators, R&D centers, and regulators also participate in this research forum. This gives WWRF an unparalleled strategic weight, not equally found in other initiatives. Given the importance of WWRF, this section will review its organization, views, and deliverables.

In a nutshell, the objectives of WWRF are (1) to generate shared visions on the research directions to follow and (2) to identify, propose, evaluate, and promote technologies likely to become the key building blocks of future wireless communications systems. WWRF can be seen as a consensus-building prestandardization initiative aimed at harmonizing views on future wireless networks. The initial convergence would pave the way to a more straightforward standardization process once spectrum is allocated by the World Radio Communication Conference (WRC) in 2007. The WWRF goals are achieved through individual and joint research activities as well as through open discussions, all centered around a number of *Working Groups* (WG). The main technical working bodies of WWRF consist of six WGs and three *Special Interest Groups* (SIGs) covering the most relevant areas of future 4G systems. Figure 5.1 illustrates the structure of WWRF and the general outline of the work carried out in each WG and SIG, indicated with a list of related white paper topics.

The work of the WWRF comes from member contributions. These contributions are then sorted, edited, and processed, in order to arrive at various deliverables. Current deliverables include:

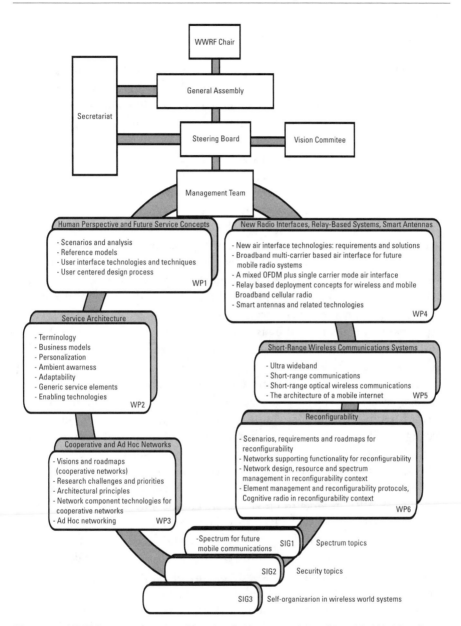

Figure 5.1 WWRF structure and main topics (white papers) considered in Working Groups and Special Interest Groups.

1. White papers and regular technical contributions on various topics, initiated by the WGs and SIGs;

2. The "Book of Visions," for general reference use and for publication [3];

3.   IEEE Magazine contributions.

Additionally, there is the all-important human networking that takes place during the regularly spaced meetings. The "Book of Visions" [3] is currently in the process of being updated and republished, and it is likely to be updated on a regular basis thereafter (in line with the further development of new 4G concepts and visions).

In an attempt to describe a user-centric wireless world, the WWRF advocates the use of a multisphere model [3] where each sphere can be seen as a particular level of interaction in the communication realm. Figure 5.2 illustrates the multisphere concept for future wireless communications systems. The innermost position in this model is occupied by the user. The surroundings closest to the user, stretching only within the reach of his or her arms and including also the user's actual body and wearable elements, define the operational environment of a *personal area network* (PAN), indicated by (1) in Figure 5.2. The user could wear, carry, and hold a number of devices that can communicate not only with each other but also with devices in other spheres of the model. The *immediate environment* (2) is formed by the familiar elements (e.g., devices with communication capabilities) located at a close distance to the user, typically within the user's room. These elements, home appliances, office tools, and so forth, would be simply, cleverly, and intuitively connected to the user-that is, in an invisible, always-present fashion. *Instant partners* (3) represent ad hoc associations with other people or complex entities (e.g., vehicles) for interacting, exchanging information, and socialization purposes. In principle, any communication taking place within or between the three first spheres can be classified as short range, owing to the fact that the maximum distances involved are on the order of 1m (1), 10m (2) and 100m (3). A number of key techniques are being studied in WWRF in order to provide short-range connectivity, based on radio (e.g., multicarrier systems, UWB) or based on optical wireless communications. The next level in the model covers a wide area and it corresponds then to the *radio access* (4) sphere. At this level full coverage is provided, including connectivity of remote users and connectivity of the lower spheres. WWRF foresees that in addition to conventional radio access technologies (e.g., base stations, satellites) a number of new access techniques will be incorporated. These include base stations in high altitude platforms (HAPS), high-speed local media points for digital broadcasting systems, and other special air interfaces. Based on the above model and definition, one can expect that the number of entities with the potential to be connected (individuals and machines) is virtually beyond measure. The next sphere models the provision of seamless wireless *interconnectivity* (5). Given the high degree of heterogeneity of terminals, air interfaces, and networks, it is likely that adaptive and reconfigurable solutions as well as radio convergence layers will be used to ensure ubiquitous and seamless operation. To this

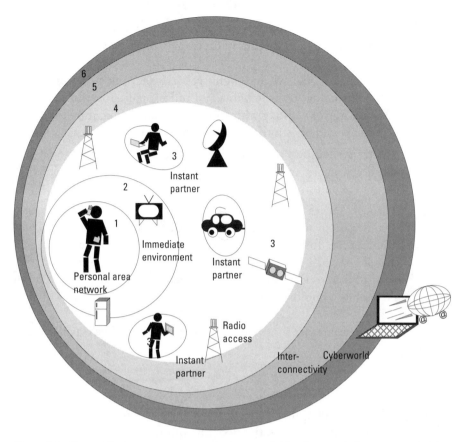

Figure 5.2 The multisphere concept, a reference model for future wireless communication systems.

end, at the highest level of abstraction, WWRF includes the *cyber world* in the model (6). This extension of the real world allows users to interact in an artificially created environment, which it manifests as realistic and appealing. Ubiquitous wireless access to the cyber world completes the overall scene envisaged for future communications systems.

WWRF has also identified a number of fundamental building blocks for future wireless communication systems. These are: (1) 4G radio interfaces, (2) cooperative networks and terminals, (3) smart antennas and base stations, (4) software defined radio, (5) semantics aware services, (6) end-to-end security and privacy, (7) heterogeneous ad hoc networking, and (8) peer discovery. The work in different WPs and SIGs is basically organized around targets defined by the sphere model and these building blocks.

As mentioned, WWRF is now the largest global initiative focused on future 4G systems. There are many possible reasons why the WWRF is enjoying high interest in the current world of wireless technology and why there was a need for this kind of initiative in terms of developing the Forum.

1. Some areas of the wireless industry were showing promising signs (WLAN coming into the home and office), indicating that a newer, different way of working with wireless technology might be emerging, one that would benefit from research investment in the area.

2. Spectrum allocation was, and still is, a lengthy process, and despite the slowdown in the mobile industry, some measures are still needed to ensure the industry does not stagnate altogether.

3. Through technical collaboration and through a united framework, better research results might be expected for future 4G mobile communications. Investment risks can be kept to a minimum in such a shared approach since collaboration reduces research costs, a fact that is greatly appreciated in times where there is an apparent recession in the mobile communication sector.

The above reasons give some clues as to why the WWRF is experiencing a high interest rate.

5.2.2 Mobile IT Forum

Founded in Japan in 2001, the Mobile Information Technology Forum (mITF) [4] is an industry-driven organization looking into future mobile communications systems. It has grown into a sizable association with about 130 members, both from Japan and elsewhere. The mITF's members profile is oriented mainly to industry, but there is some participation of network operators and service providers. Connection to the local academia is provided through individual members-professors affiliated to key Japanese universities. The core activities are concentrated around the Fourth Generation Mobile Communications Committee and the Mobile Commerce Committee. Figure 5.3 shows the structural organization of the mITF, highlighting the goals and activities of the 4G Mobile Communications Committee [4].

The mITF carries out research and development activities focused on essential aspects of future 4G systems. So far, the most important outcome from mITF is the "Flying Carpet" report [5]. This document approaches future 4G systems from user behavior, society, and industry standpoints. It considers feasible scenarios and features of future systems that match user expectations. Requirements and technical challenges for future systems are identified from the service platform and system infrastructure perspectives. The "Flying Carpet" report also proposes feature and technology roadmaps as well as reference models and concrete proposals for 4G mobile systems. As with the "Book of

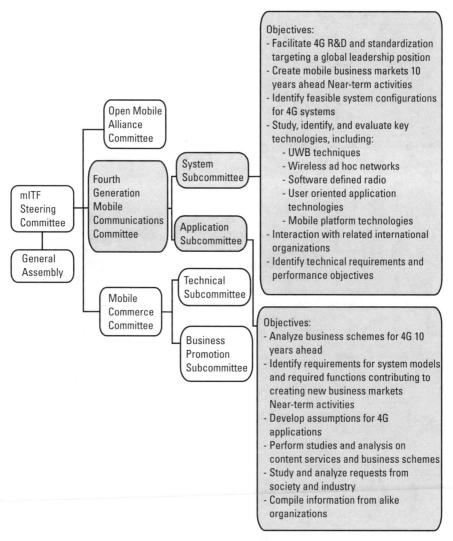

Figure 5.3 Structure of the mITF.

Visions," a user-centric approach is adopted here too. In addition, the report targets man-to-man, man-to-machine, and machine-to-machine communications. A concise summary of the "Flying Carpet" outcomes is now presented. In an attempt to create new visions on future user scenarios, some promising settings or scenes were identified in this report. These are shown in Figure 5.4

Based on the above scenarios, one can see that 4G mobile systems will be a part of everyday life. User surveys showed a high degree of acceptability to the use of wireless communication systems in these scenarios. As for user's expectations, the illustration in Figure 5.5 crystallizes the results of the surveys.

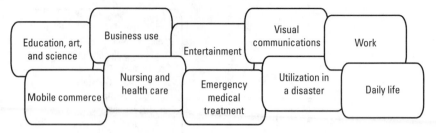

Figure 5.4 The most likely scenarios of application of 4G mobile systems according to mITF.

- Freedom in time, space, and functionality
- Need to have more natural communications
- Reasonably low cost of terminals and services
- Bandwidth and fun-factor or convenience-factor not necessarily proportional
- Intermediate agent between user and information banks to rationalize retrieval operations
- Mere high-speed is not enough; something more appealing is desired

User expectations

Figure 5.5 User's expectations on future 4G mobile systems.

Infrastructure
- Transmission rates of 100 Mbps or higher
- Stable and secure connection in critical applications (medical, e-commerce, etc.)
- Bandwidth dedicated for public service
Terminals
- Highly heterogeneous terminal range requires interoperability and compatibility
- Sophisticated features with affordable prices
- Attractive and effective media terminals
- Easy-to-use universal user interface

Industry expectations

Figure 5.6 Industry expectations on future 4G mobile systems.

The study also investigates the industry expectations on future 4G mobile systems. These results, obtained based on applications and business models also developed in the mITF, are summarized in Figure 5.6.

From the point of view of the required quality for 4G systems, it is agreed that data rate is only one among several important desired characteristics. The most relevant quality indicators are the following: (1) sufficient transmission rate (e.g., 100 Mbps in movement, 1 Gbps in fixed/nomadic cases), (2) certainty and stability of connection, (3) multipoint, interactive real-time communications, (4) provision of flexible communications for diversified service usage, (5) highly secure communications, (6) sufficient storage capacity in terminals (e.g., 100 GB), and (7) low power consumption and high-capacity batteries.

In order to implement future 4G systems the most appropriate or promising technologies have to be identified and further evaluated to validate their suitability. The Forum has identified a number of key techniques and technologies with high potential to become essential building blocks for 4G systems. These techniques and technologies, classified into four major categories, are shown below.

1. High-speed and large-capacity wireless transmission techniques:

   - *Frequency refarming:* Rearrangement of allocated frequency bands to allow more efficient spectrum utilization and study of new bands.
   - *Advanced adaptive techniques to increase spectral efficiency:* Adaptive modulation and coding, adaptive control for automatic repeat request (ARQ), adaptive array antennas.
   - *MIMO techniques for exploiting spatial multiplexing:* Fundamental for attaining very high data throughput.
   - *Multicarrier techniques:* Frequency orthogonal techniques (e.g., OFDM) as the baseline modulation technique for 4G systems-used in general in combination with CDMA and TDMA techniques give more design flexibility and interference protection.
   - *Interference and fading mitigating techniques:* These two major impairments in typical wireless scenarios may cause serious de-gradation to systems performance. Interference cancellation, equalization, and multiantenna techniques, among others, are key approaches to be investigated.
   - *Error control techniques:* Advanced FEC and ARQ schemes need to be incorporated to 4G systems to guarantee a desired signal quality regardless of the channel conditions.
   - *High-speed packet radio:* Important techniques needed to improve the utilization of radio resources, allowing higher system capacity and higher data rates.
   - *Handover techniques:* As seamless communications is one of the most distinctive features of future systems, fast and reliable

handover techniques for a considerable large number of scenarios should be investigated.

2. Network technologies:

- *Radio access networking techniques:* These techniques will enable the dynamic establishment of multiple continuous or isolated access cells.

- *Robust networks:* Variable-capacity networks able to adapt to the variable traffic demand conditions while keeping performance unchanged is a challenging area for research. Multiple antennas, congestion control, and multihop communications have been identified as promising techniques for this purpose.

- *Ad hoc networks:* Rather than the centralized approach of conventional wireless communications where communications use a central node (base station) to access the network, an ad hoc network can be seen as a distributed approach where the signal can hop through several nodes (terminals) to reach the destination. Routing protocols, user and service discovery procedures, and security issues are the main research items.

- *Seamless networking techniques:* Allows seamless interconnection of a packet-based backbone network with radio access or other networks.

- *Approach link techniques:* A number of techniques used to connect base stations or remote stations to control stations by using optical means (fiber, free-space optics), microwaves, and so forth.

- *High-speed transport technology:* Techniques to transport and switch efficiently very high-speed packet data.

3. Mobile terminal technologies:

- *Circuit and component technology:* Faster processing circuits, denser integration capabilities, smaller batteries, and display technologies with better image quality are identified as important goals to be fulfilled to fully comply with 4G system requirements and user expectations.

- *Battery technology:* Enhancement of battery capacity, efficiency, operational time, and size is crucial to succeed in producing future handheld terminals. Low-cost batteries and environmentally friendly materials are also very important facts to consider.

- *Human interface:* This is recognized to be a very important research area, where natural ways of interaction between user and machine are of the utmost importance in order to make future 4G systems attractive. Terminal form factor, wearable terminals, visual and voice interfaces, and the use of biometric information are among the most important research issues.

- *Terminal security techniques:* Techniques aiming to provide levels of security at least similar to those found in networks using public key-based infrastructure should be developed. User *authentication and data protection are also highly relevant research topics.*

- *Terminal software:* Common operating systems, software platforms, and applications compatible with SDR should be developed.

- *Multisystem wireless terminal technologies:* It is envisaged that future terminals will be able to handle seamlessly short-range, wide-area, and broadcast communications using different technologies implemented in the terminal. There is a series of technical challenges that need to be solved, including multisystem radio for multiple modulations, protocols and frequencies, multiband antennas, practical realization of SDR, and seamless operation between the systems.

- *Software defined radio:* Fundamental technology applicable in both base stations and mobile stations permitting the use of multiple access technologies in a very efficient fashion. Issues to be studied include transceiver architecture, basic implementation technologies, and basic transmitter and receiver technologies realizable in software, e.g., multimode modulation and demodulation techniques, multiband transmit and receive techniques, and so forth.

4. Mobile systems technologies:

- *Quality of service:* These techniques encompass different methods to control and ensure fundamental communication parameters, like desired transmission speed, bandwidth, delay, and packet loss rate.

- *Mobility control:* Techniques aiming to provide and maintain QoS in environments characterized by mobility. These include handovers and mobile Internet protocols.

- *Mobile multicast techniques:* Broadband transmission techniques for real-time and nonreal-time broadcasting of information.

- *Location determination and navigation:* Reliable techniques for providing user coordinates, speed, and other navigation information in outdoor and indoor scenarios. A large number of applications could potentially exploit this information.

- *Security, encryption, and authentication:* Techniques aiming to secure mobile communications by preventing any type of illegal use of the wireless resources.

5.2.3   Future Technology for Universal Radio Environment Project

The Future Technology for Universal Radio Environment (FuTURE) Project [6, 7] is a government-driven research project in China, part of a major research initiative known as the National High Technology Research and Development Program (or, for short, the 863 Program). It is organized in plans of 5 years length, and FuTURE corresponds to the 2001 to 2005 period. Extensions of the project are already planned. Figure 5.7 shows the plans and goals of the FuTURE Project until 2010, where three planned phases are shown. The target of FuTURE is the development of concrete solutions for future 4G systems. This is done by developing some demonstrative experiments supported by research activities. The long-term goal of the project is to put China in a competitive R&D position once standardization and development activities on 4G are started on a global scale.

Members participating in the FuTURE initiative include local R&D units and universities. In addition, an important number of foreign manufacturers take part in the project through local affiliates. The FuTURE Project has liaisons with WWRF, the European Sixth Framework Program (FP6), mITF, NGMC (Korea), and other 4G related programs.

The FuTURE Project advocates the use of multiple wireless communications layers, all connected through an IPv6-based core network. At the upper

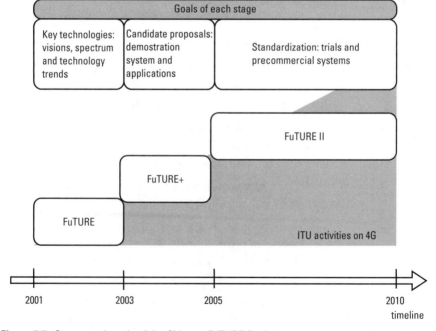

Figure 5.7 Stages and goals of the Chinese FuTURE Project

level the broadcast layer provides access through a HAPS situated at 10- to 30km altitude. Wide coverage is provided by the cellular layer, seen as an extension and evolution of 3G systems into B3G/4G systems. Finally, area layers provide local coverage access, including wireless local, home, and personal area networks (WLAN, WHAN, WPAN).

The FuTURE Project is organized into the following subprojects:

- B3G Radio Access Techniques;
- Wireless LAN and Ad Hoc Networks;
- Multiple Antenna Environment (MIMO) and RF;
- 3G-Based Ad Hoc Networking;
- IPv6-Based Mobile Core Networks;
- Generic Techniques for Mobile Communications;
- System Structure, Requirement, and Higher Layer Applications.

The general requirements envisioned by FuTURE for future wireless systems (B3G/4G) are as follows: (1) large dynamic range in data rate, from 10 Kbps to 100 Mbps, (2) high spectral efficiency, (3) packet data as the dominant service, (4) universal use of IP addressing (IP assigned to every possible entity), and (5) low power transmissions and strict compliance of EMC norms.

The project has also identified a number of important research topics and challenges key to the development of future wireless systems. These can be summarized as follows: (1) efficient use of space, time, and frequency resources is paramount to achieving targeted performance figures, in particular high data rates, large system capacity, and wide coverage; (2) MIMO systems allow the attainment of large spectral efficiencies (2 to 5 bps/Hz and higher) but practical implementation is not straightforward; (3) time division techniques complement well with packet data traffic; (4) space-time coding techniques; (5) development of turbo receiver concepts; (6) TDMA appears to be an attractive multiple access scheme, possibly used in combination with CDMA and FDMA elements; (7) link adaptation techniques, including adaptive modulation and coding and ARQ schemes; and (8) novel channel coding techniques. Two parallel developments are being conducted, differentiated broadly by the employed duplexing scheme. These are organized and managed separately by two different groups of local universities. The time division duplexing (TDD) system considers a TD-MC-CDMA system able to serve both wide and local areas. Target peak data rates are 30 to 50 Mbps and 40 to 100 Mbps, respectively. The frequency division multiplexing (FDD) system, based on generalized multicarrier/OFDM techniques, aims to provide wide coverage access with peak data rates of 30 to 50 Mbps.

5.2.4   Next Generation Mobile Communication Forum

Research activities on B3G systems started in Korea in 2000, fueled initially by the Ministry of Information and Communication (MIC). In 2002 the 4G Mobile Communications Vision Committee was established. At an early stage, the government-funded Electronics and Telecommunications Research Institute (ETRI) was the instrumental organization coordinating 4G research activities in Korea. Today, participating organizations include local and international manufacturers, national universities and operators, and ETRI. In 2003 this Vision Committee was reorganized and structured as a forum, the Next Generation Mobile Communication (NGMC) Forum. A considerable amount of activities related to future 4G systems are carried out in Korea today. From these, probably the most ambitious in terms of scope, size, and international projection is the NGMC Forum [8]. Most of the scope and goals of the NGMC are well in line with those of the WWRF, mITF, and FuTURE initiatives. The most important targets are as follows: (1) analysis of technical and social trends, (2) vision establishment, (3) standardization and international cooperation, (4) advanced R&D strategies steering, and (5) spectrum allocation planning. Figure 5.8 summarizes the NGMC Forum activities and organization. The research activities are carried out by three WGs. Their goals are briefly presented below.

1. Market and Service Working Group (WG):
   - Analyze trends of mobile communication markets and services.
   - Propose policies for the activation of mobile Internet markets.
   - Develop new services in fixed and mobile convergence (FMC) environments.
   - Analyze broadband services trends and practical possibilities in wireless environments.
   - Forecast 4G demands from 3G and mobile Internet markets analysis.

2. System and Technology Working Group:
   - Define and select 4G technologies.
   - Valuate 4G technologies and set up technical goals.
   - Propose technical standardizations in 4G.
   - Cooperate with international forums for technical developments.

3. Spectrum Working Group:
   - Analyze trends of spectrum utilizations in mobile communications.
   - Propose spectrum utilization plans for the activation of mobile communications.
   - Discuss 4G spectrum.

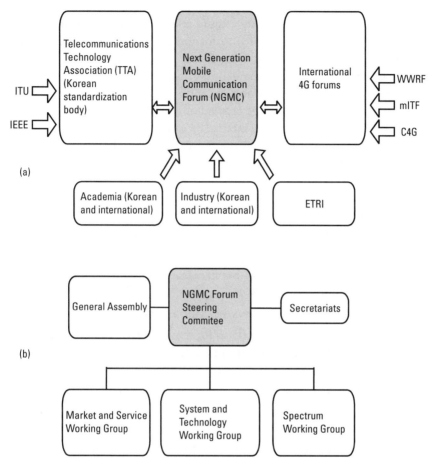

Figure 5.8 (a) NGMC Forum interaction with domestic and international organizations, and (b) NGMC local organization.

5.2.5 4G Research Cooperation Projects in the European Sixth Framework Program

EU-funded projects focused on 4G research are presently organized around the Sixth Framework Program (FP6) initiative. The FP6 sets out the priorities for research, technological development, and demonstration activities for the period 2003 to 2006. The EU-Information Society Technology FP6 has two main strategic objectives:

1. Strengthen the scientific and technological bases of industry.
2. Encourage its international competitiveness while promoting research activities in support of other EU policies.

There are a number of activity areas of relevance to the Integrated Projects (IP) that comes under the title of Information Society Technologies. An IP is regarded as an instrument of the FP6. It is a multipartner project to support objective-driven research. The main output/deliverable expected of an IP is knowledge for new products, processes, and services. The goals of an IP are expected to be ambitious and to be focused on increasing Europe's competitiveness or addressing major societal needs. In order to achieve these ambitious goals, IPs are expected to bring together a large number of partners in order to reach a critical mass. A nonexhaustive list of IPs is given below:

1. My Personal Adaptive Global Network (MAGNET) (http://www.ist-magnet.org);
2. Wireless World Initiative New Radio (WINNER) (http://www.ist-winner.org);
3. Mobile Applications and Services Based on Satellite and Terrestrial Interworking (MAESTRO);
4. Ambient Networks (http://www.ambient-networks.org);
5. Designing Advanced Interfaces for the Delivery and Administration of Location Independent Optimized Personal Services (DAIDALOS) (http://www.ist-daidalos.org/main.htm);
6. IP-Based Networks, Services, and Terminals for Convergent Systems (INSTINCT);
7. Multimedia Networking (MEDIANET);
8. End-to-End Reconfigurability (E2R) (http://www.e2r.motlabs.com);
9. Pervasive Ultra-Wideband Low Spectral Energy Radio Systems (PULSERS) (http://www.pulsers.net).

Two large-scale, broad-scope EU projects are presented below, namely MAGNET and WINNER. These research projects comprehensively explore the two main components of 4G systems: local access (MAGNET and WINNER) and wide coverage access (WINNER). Both projects were started in January 2004. Each of the projects has nearly 40 participating organizations-essentially manufacturers, academia, research laboratories, and small and middle size enterprises. Partners are predominantly from Europe, but participation from Asia and North America is reasonably high.

5.2.5.1 MAGNET Project

The MAGNET project [9], coordinated by the University of Aalborg, Denmark, is aimed at the development of a new concept based on a user-centric personal communications system. The concept revolves around a *personal network*

(PN), which supports all of a user's activities (at home, in the office, away on business) without being obtrusive and ensures that services are delivered in a reliable, secure manner and with good quality of service. A key element of the PN is a PAN also referred to as a personal bubble or "the user's immediate spatial environment." PANs are linked to each other by (yet to be defined) structures, forming a user's PN. Devices that do not belong to a user specifically are treated as foreign devices and these may also, under the right conditions, be connected into a user's PAN (and thereby to the user's PN).

The PNs and PANs are reconfigurable due to their very nature, with configuration and connectivity depending on resource availability, time, place, preference, and context.

The MAGNET project does not simply address appropriate radio technology, it is a complete system design taking into consideration scenarios, business models, PAN architectures, PN architectures, networking and internetworking at both the PAN and PN level, security and privacy, and radio access technologies appropriate for PANs.

Six work packages (WPs) cover the research area. A brief presentation of each WP is included below.

- WP1: User Requirements and Socio-Economic Aspects. In this WP user requirements and scenarios for PN are considered and proposed, based on user-centric approaches.

- WP2: Networks and Interworking. This vast WP explores and proposes the following network issues: (1) the role and impact of IPv6 in implementing the PN, (2) requirements for PN architecture, (3) resource and service discovery techniques, (4) context discovery techniques, (5) ad hoc self-organizing and routing architectures, and (5) requirements and solutions for naming and addressing in PNs. In addition, PAN interfaces with infrastructure and active network support for PNs are also considered.

- WP3: Adaptive and Scalable Air Interfaces. Adaptive and highly spectra efficient air interface approaches are investigated in this WP. Multicarrier and UWB techniques are considered as the baseline candidates for short-range physical layer. In order to evaluate the considered schemes, characterization and modeling of the PAN channel are required.

- WP4: Security and Privacy Issues. Efficient approaches for providing security and privacy to future PANs are investigated in this WP.

- WP5: Flexible Platforms and Prototypes. In order to build a demonstrator at the final stage of the project, software and hardware

implementation approaches are studied in this WP, in close cooperation with WP3.

- WP6: Dissemination and Standardization. This WP takes care of
dissemination and standardization activities, including development of
the project Web page, cooperation with peer projects and standardization bodies, market analysis, intellectual property rights policy, and so
forth.

5.2.5.2 WINNER Project

WINNER [10] is a research consortium coordinated by Siemens and focused on
enhancing the performance of mobile communications systems. The WINNER
project is aimed at the development of a new concept when looking, in particular, at the radio access part of mobile communications systems. This concept
evolves around the expectation that communication systems of the future will
converge, which in turn may lead to one global system and fewer unique systems. The project also looks to address the concept of the *ubiquitous environment*, where technology is embedded around users, enabling them to access
information anytime, anyplace, and anywhere, using any device. This is an
ambitious goal, particularly if legacy systems are to be taken into account.
WINNER will look into radio access adaptability research, where adaptability is
seen in the context of multiple scenario adaptation. This implies that the system
to be investigated is expected to work across a wide range of environments (short
range to long range) and will have the ability to provide multiple services and
applications to all users.

In line with the ability to offer increased services and enhanced quality of
service, WINNER seeks to examine radio access technology for achieving:

- An increase in peak data rate;
- Less delay (improved latency);
- The ability to cope with higher mobile speeds;
- More efficient spectrum utilization;
- Improved coverage;
- Significant improvements in cost per bit (to all mobile players).

Concepts towards achieving the above include the following:

- Utilizing advanced and flexible network topologies;
- Utilizing advanced and flexible physical layer topologies;
- Making use of efficient frequency sharing methods.

All the above is to be done in a fairly autonomous fashion since the expectation is that the user will not care about the underlying technology.

Two key environments have been identified for the first project phase: (1) short range, and (2) wide area. Characteristics pertaining to the short-range environment include high cell throughput (up to 1 Gbps, indoor and outdoor) and use of multihop concepts to improve coverage and system scalability. Characteristics pertaining to the wide-area environment include peak data rates per user of 1 to 100 Mbps, usage in urban, suburban, and rural areas, and the ability to cope with high mobility.

There are seven operational WPs in WINNER. A brief description of these WPs follows.

- WP1: Scenarios. In this WP the user scenarios are defined regardless of any technical consideration. Initial working assumptions are obtained from the different partners. Key motivations and user categories are then used to define the scenario elements.

- WP2: Radio Interface. The radio interface is one of the main aspects of the WINNER concept, and it is comprised of leading-edge technologies for the physical layer, MAC, radio/data link control, and radio resource management. The air interface is to be capable of working in several different modes. Some important assumptions for the physical layer are as follows:

  - Multicarrier transmission is assumed for the downlink;
  - For the uplink several options are to be researched;
  - With duplex schemes three main areas are to be researched: TDD; FDD, and hybrid FDD/TDD.

- WP3: Radio Network Deployment Concepts. In line with the research into a new radio interface is the related area of radio access network research and deployment concepts. The research involves comparison with standard methods to arrive at a concept that enhances capacity and performance, cost, complexity, spectrum efficiency, related EMC properties, and EMF strength. Research in this WP includes mobile and cooperative relaying/distributed antenna systems.

- WP4: Cooperation of Radio Access Systems. Traditional cellular systems employ but one radio access technology-even if they ultimately link back to common supporting infrastructure or make use of shared infrastructure. Research in this WP goes towards a more integrated approach in terms of providing cooperation between different radio access technologies. Research includes defining cooperative mechanisms between WINNER radio access networks and legacy networks, identification of information required for handover and radio resource management, and the impact of cooperative mechanisms (on air interface, architecture, and protocols).

- WP5: Channel Modeling. Analyzing and selecting existing channel models, determination of propagation scenarios, and channel measurements and modeling are the main tasks of this WP.

- WP6: Spectrum and Coexistence. Without the availability of spectrum for a system, its future deployment would be difficult to attain. It is therefore necessary to address a methodology for calculating spectrum requirements for input to spectrum licensing bodies. This WP participates and contributes in ITU-R WP 8F. Study items include spectrum range requirements, methodology for estimating spectrum requirements, requirements for extending existing (e.g., 3G, B3G) systems, efficient sharing methods, flexible spectrum use methods, coexistence, and contributions to the regulatory process.

- WP7: System Engineering. Designing any one aspect of a communication system in isolation can be a big mistake. Ensuring that a system's component modules can be integrated throughout the research, design, build, and test lifecycle is the task of this WP.

5.2.6 The Worldwide Wireless Initiative

The Worldwide Wireless Initiative (WWI) [11] is a joint effort that gathers more than 100 partners from industry, research laboratories, academia, and regulatory agencies. This chiefly European initiative has extended its operational area globally, now incorporating members from America, Asia, and Oceania. WWI contains a series of large Integrated Projects in the FP6 Information Society Technology. The WWI takes a further step towards developing essential technologies for 4G by creating links among important European projects with the aim to produce highly relevant research proposals and concrete results. The vision of ambient intelligence around the user is advocated by the WWI, where the needs, expectations, and requirements of the user will be supported at all system levels. One of the objectives of WWI is to define systems and functions that provide the users with the best possible experience at the lowest possible cost of equipment, use, and maintenance.

Research activities were started at the beginning of 2004 and they are planned to span a 6-year period, divided into three 2-year phases. Phase I deals with the initial steps of identifying the baseline technologies and defining the requirements. Three FP6 project will cover this phase: WINNER (new radio interfaces) [10], Ambient Networks (networking) [12], and E2R (end-to-end reconfigurability) [13]. Technology development as well as system definition will be considered by Phase II, while Phase III will deal with system synthesis and demonstrations. WWI intends to contribute to standardization activities and regulatory bodies as well.

5.2.7 Samsung 4G Forum

In an attempt to promote the development of 4G technologies worldwide and address the need for creating a consensus among key players, Samsung established the Samsung 4G Forum in 2003. The driving force behind this forum is the strong belief that industry, operators, academia, and regulatory bodies need to establish a fluid dialog to build sound foundations for future 4G systems through a better mutual understanding. This annual event gathers leading experts from all over the globe to discuss strategic and technical aspects of 4G systems. Every year the forum carries a main theme, reflecting major current issues being considered by the related community. In year 2003 the forum theme was "Global Strategies and New Air Interface Technologies," whereas in 2004 the main subject of the forum was "Migration Paths Towards 4G Networks." In 2005 the theme will be "(R)evolution of Radio Access Technology over All-IP Based Networks." The forum has shown a unique ability to congregate key people from companies, universities, and other 4G related parties to exchange their views and strategies. A brief list follows of important issues that appear to be present in most if not all the visions.

- 4G should be developed around the user and it should not be driven by technology. It should be a user-centric approach where technology is critical but not the goal.

- 4G cannot be developed in isolation-legacy generations need to be taken into account.

- 4G is synonymous with always being connected, anywhere, at any time, with seamless operation.

- The success factors for 4G are services and applications, research in technology, strategic alliances, and standardization.

- Reductions in cost will give 4G the needed thrust for success; cost per bit and cost of terminals and infrastructure should all be reduced.

- Several new services and application models developed around the foreseen capabilities of future 4G systems show a huge potential for such systems to become widely used.

- The task of fulfilling simultaneously all 4G requirements is far from trivial.

- 4G could be defined with evolutionary and revolutionary components.

- 4G is a convergence platform: devices, networks, services, and information will converge.

- 4G solutions are approached through new radio technologies and evolution of cellular and local access systems.

- A joint common definition of 4G and related terminology is missing.

- 4G networks will be highly heterogeneous.

- Battery technology, particularly power efficiency and capacity/volume, is not evolving as future power hungry devices and applications.

5.2.8 The eMobility Technology Platform

The eMobility Technology Platform [14] is a large-scale initiative aimed to strengthen the European leadership position demonstrated in mobile communication systems (e.g., 2G and 3G systems), paving the way to a similar success in future communication systems. Developing effective technical solutions to cover the whole range of scenarios and capabilities requires an open dialog and extensive collaboration between involved parties. Major European telecommunication manufacturers and operators are the main actors behind eMobility. The platform will focus not only on technical solutions but also the on creating a rich environment for developing new services and applications. This initiative can be seen as a European response to the recent successful developments in mobile and wireless communications witnessed in Asia and North America. Europe is planning to invest heavily in order to secure a key position in future communication systems. This is demonstrated by the budget of 2 billion in a period of 2 years, an unparalleled amount by comparable initiatives. The Strategic Research Agenda [14] was recently published, offering visions on user experience and acceptance, business infrastructure, services, network and radio technology, security, contents, and so on. The eMobility technology platform envisions a highly integrated and homogeneous network of networks, including wideband wide area systems, fixed wireless access, local access (e.g., WLANs), and personal wireless communications (e.g., PAN and BAN).

5.2.9 Other 4G Research Initiatives

The list of projects presented above, though not exhaustive, embraces key 4G projects. A number of other important projects are briefly presented in this section. The 4G Mobile Forum (4GMF) [15] is an initiative originating in China but with increasing international presence. WWRF can be seen as a research-oriented initiative, while the 4GMF defines itself as a development-oriented forum. The 4GMF is encouraging the development of 4G mobile communications with the so-called Open Wireless Architecture (OWA).

North America is also active in 4G research activities, although, in general, activities are divided into smaller research projects, each project generally involving a single university or company. IEEE standardization and development activities towards different air interfaces (e.g., IEEE 802 activities) will be discussed in the next section. In the United States the Defense Advanced Research Projects Agency (DARPA) supports an approach of flexible spectrum allocation. Its neXt Generation (XG) Communications program [16] is developing techniques to allow multiple users to share the spectrum through adaptive mechanisms that resolve conflicts among users in time, frequency, code, and other domains. DARPA's goals are to enable a tenfold increase in spectral efficiency. The DARPA program develops technology that is applicable not only to the military but also to civil use as well. The Project Oxygen [17] is a 5-year alliance between the Laboratory for Computer Science (LCS) and Artificial Intelligence Laboratory (AIL) at the Massachusetts Institute of Technology (MIT) and a number of major industry leaders. The vision of the project is that in the future, computation and communications will be human centered, and that they shall be freely available everywhere, at any time, like the oxygen in the air. The project identifies a number of high-level challenges for a future communication system. Thus, to support human activities in a natural way, future communication systems should be *pervasive* (available everywhere), *embedded* in our world, *nomadic, adaptable* to user and operating conditions, *powerful* (yet efficient), *intentional* (e.g., intuitively close to the user), and *eternal* (available at any time). A countless number of research projects on 4G related technologies are currently active in companies, universities, and research institutes across North America. These projects, concentrating on essential techniques for future communications systems, are conducted independently, without aiming at consensus building among the research institutions. In that respect, the same trends are found all over Europe and in several Asian nations, where innumerable autonomous projects investigate enabling technologies for future communication systems, in addition to the large-scale initiatives discussed previously in this chapter. In Europe in particular, there are several nationwide research programs with strong affinity toward research in several technical areas of 4G systems. Examples of such initiatives are the FUTURA Project in Finland (Future Radio Access) [18] and the Virtual Centre of Excellence in Mobile and Personal Communications (MVCE) in the United Kingdom [19].

5.3 Paving the Way to 4G: Worldwide Development

In the previous section several 4G research initiatives were introduced and discussed. In parallel to these research oriented activities, development work is currently being carried out by several parties. Such work, though tightly coupled to research activities, is more focused on the development and implementation of concrete practical solutions capable of demonstrating particular features of future high-performance systems. It is not the intention of this section to present a comprehensive overview of all the development work in progress, but rather to introduce a few representative works.

5.3.1 NTT DoCoMo (Japan)

DoCoMo has been championing the use of a single radio solution for all environments. This would certainly bring significant advantages to manufacturers, operators, and end users. In fact, the DoCoMo system supports both cellular and hotspot access using one air interface with different radio parameters. Frequency bands below 6 GHz and coverage areas with 500 to 700m cell radius in urban areas and 2 km in suburban or rural areas are being considered. System requirements also include support of around 300-km/h mobility targeting high-speed public transportation, such as bullet-trains. Regarding system capacity, twice the system spectral efficiency of 3G systems (1 bps/Hz) is required without advanced technologies, such as MIMO spatial multiplexing. System bandwidths for downlink and uplink have been assigned asymmetrically. The wireless access techniques adopted for each direction of communication are also different. As will be described in more detail, downlink employs VSF-OFCDM, while uplink uses variable spreading and chip repetition factor CDMA (VSCRF-CDMA). Both CDMA and TDMA are considered as intersite, intersector, and intrasector multiple accesses, even though the details are not specified yet.

The key feature of VSF-OFCDM is the spreading over the two-dimensional time-frequency domain. The primary purpose of spreading is to obtain frequency diversity. Due to the very large signal bandwidth (100 MHz) compared with the coherence bandwidth in typical conditions, frequency domain scheduling would require large signal bandwidth for uplink control signals. This is because in order to properly address (via feedback) a large amount of subcarriers in a wide bandwidth, the system needs a considerable number of (overhead) bits. For that reason the system exploits frequency diversity (e.g., free of feedback control signals) rather than using frequency domain scheduling. The spreading also offers randomization/mitigation of intercell interference. Other merits of using code division multiplexing (CDM) features include, in their terms, easy support of low rate physical channels and flexible resource allocation by simply changing the spreading factors.

For uplink transmission, the VSCRF-CDMA works as follows: First, the input sequence, which would be a coded and symbol mapped complex sequence, is spread with channelization code and scrambled with a long code as usually done in CDMA. Then, a block of chip sequences is repeated such that the output has comb-shaped frequency spectrum. Finally, this output is modulated with a user-specific phase sequence, for which a set of DFT sequences is commonly used to ensure orthogonality between users. Spreading factor and chip repetition factor are adaptively changed according to the number of simultaneous access users and radio channel conditions. Originally, the scheme was first introduced in [20] without considering the prespreading.

5.3.2 Samsung's Terrestrial OFDM Packet Access System

Recently, Samsung proposed a terrestrial OFDMA-based packet access system. The system assumes 2- to 5-GHz frequency band and a scalable system bandwidth with unit bandwidth of 20 MHz. About 60-Mbps peak data rate can be achieved in 20-MHz bandwidth without MIMO spatial multiplexing. The system requirement includes higher spectral efficiency than twice the system capacity of 3G systems (i.e., higher than 1 bps/Hz with frequency reuse of 1) and support of high-speed vehicles, such as KTX, a Korean high-speed rail service with a speed greater than 250 km/h). System bandwidths for downlink and uplink have been assigned symmetrically.

One of the merits of using OFDMA is the multiuser diversity gain that can be obtained by means of frequency domain scheduling. This frequency domain scheduling could be considered to be one of key features that differentiate the system from the existing 3G wireless systems based on spread spectrum techniques that are mainly targeting frequency diversity. For downlink transmission, the system defines a frame cell (FC) as a finite time-frequency resource consisting of M consecutive subcarriers and N OFDM symbols, to be used as a unit for channel quality feedback, and adaptive modulation and coding. An FC is further divided into multiple subchannels and can be shared by many users. For this purpose, the system specification also defines a time-frequency cell (TFC) consisting of a set of adjacent subcarriers, where the number of subcarriers belonging to a TFC is smaller than M. Moreover, a subchannel is defined as a sequence of hopping TFCs. With this type of structure, the multiuser diversity gain can be traded off with frequency diversity by changing the width of an FC. Note that the frequency domain scheduling requires knowledge of the channel quality for each subchannel, and it raises another trade-off between the throughput and the uplink signaling cost. Background study, however, has shown that channel quality feedback for only a few best subchannels are enough to obtain considerable multiuser diversity gain, especially when the number of users in the system is large enough. Note also that the CDM feature is also used within a TFC for interference randomization and mitigation when the system is partially loaded.

For the uplink, a similar structure is used, but without defining an FC, which is equivalent to using only one FC, and the TFCs that comprise a subchannel hop over the entire bandwidth. This is basically done to obtain frequency diversity only, since in the uplink there is no common pilot channel available. Because each user will use only a portion of the subcarriers, the peak-to-average power ratio (PAPR) problem in the uplink would be less critical than in the downlink transmitter.

5.3.3   The Wireless Broadband Project

Korea has long enjoyed a very high penetration in mobile communications as well as in (wired) wideband Internet usage, with approximate rates of 85% and 60%, respectively (2004). This has motivated national regulatory bodies, research institutes, manufacturers, and operators to develop a system and create services able to bring to the user, to a limited extent, the best of both worlds. The approach is very appealing, not aiming at extreme mobility or very high data speeds, but instead offering moderate mobility and reasonably high data speeds at relatively low cost. This project started in 2003 and it was initially known as the High-Speed Portable Internet (HPi). Nowadays it bears the name of WiBro (for Wireless Broadband) and it can be considered as a 4G precursor, covering an important midway market segment. When fully operational in 2006, this mobile Internet system will be strategically positioned, complementing both the high-speeds and low-mobility of WLAN systems and the high-mobility and low-to-moderate data speeds of cellular systems (e.g., IMT-2000). Figure 5.9 shows the approximate operating boundaries of WiBro in comparison to other wired and wireless systems.

Briefly, WiBro is defined as a wireless system characterized by moderate mobility (60 km/h), moderate data throughput (3/1 Mbps peak in DL/UL, respectively) available at any time, anywhere, at low cost. The Ministry of Information and Communications has reallocated the 2.3-GHz band, once reserved for FWA, to the WiBro initiative.

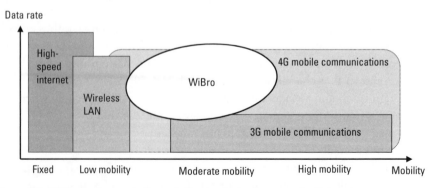

Figure 5.9 WiBro system and other wireless and wired communication systems.

Two working groups are concentrated on most of the technical and services aspects. The Service and Network Working Group deals with the definition of the network and services as well as the definition of the network reference model. The role of the Radio Access Working Group is to define the system parameters and to identify key technical items and the criteria and conditions for their evaluation. In addition, this group will evaluate the candidate proposals. Two ad hoc groups also support the activities of the WiBro project: the International Coordination and IPR groups. WiBro has established liaisons with the IEEE 802.16 [21] to cooperate in different aspects of the air interface design and standardization.

Key system parameters and requirements have been fixed since January 2004, and they are summarized in Tables 5.1 and 5.2.

The development work of the WiBro system is planned to be carried out between 3Q 2004 and 2Q 2005. After some field trials in the second half of 2005, the service is scheduled to be introduced in the Korean market in the beginning of 2006.

In addition to fulfilling local market needs, the WiBro initiative has an important and strategic value, as it will bridge the two extreme segments of 4G. In terms of penetration, WiBro could gain a large amount of the local wireless data market share, as it aims for data rates and mobility figures high enough to appeal ordinary users. It is expected that by 2010 there will be more than 10 million WiBro users in Korea.

The WiBro system is based on 802.16e [21], especially, with OFDMA mode where a set of subcarriers is defined as a subchannel to be used as a unit for resource allocation and user multiplexing. Basically, the system is a TDD system that easily provides flexible link capacity adjustment for uplink-downlink asymmetry in traffic load. Two different subchannel structures are defined in the WiBro system. A regular subchannel is defined as a set of subcarriers distributed over the entire bandwidth, dubbed the diversity channel, so that one can obtain frequency diversity; while an AMC subchannel is defined as a set of consecutive subcarriers primarily targeting to obtain multiuser diversity by means of frequency domain scheduling, much as in the Samsung's terrestrial OFDM-based packet access system described in Section 5.3.2.

Table 5.1

Key System Parameters of WiBro

Duplexing	TDD
Multiple access	OFDMA
Bandwidth	10 MHz

Table 5.2
Radio Access and System Requirements of WiBro

Service coverage	1 km		
Mobility	60 km/h		
Data throughput per user	Max.	DL	3 Mbps
		UL	1 Mbps
	Min.	DL	512 Kbps
		UL	128 Kbps
Frequency reuse factor	1		
Spectrum efficiency	Max.	DL	6 bps/Hz/cell
		UL	2 bps/Hz/cell
	Min.	DL	2 bps/Hz/cell
		UL	1 bps/Hz/cell
Handoff latency time between cell, sectors, and frequencies	≤ 150 ms		
QoS	Support of best effort, real-time polling, nonreal-time polling		
Multiple antennas	Optional support to increase coverage/through-put		

5.3.4   The IEEE 802 Wireless Standards

IEEE started to develop standards for wired networks (e.g., LAN, MAN) in 1980, under the generic denomination IEEE 802 standard. This effort evolved from developments for wired networks (e.g., coaxial cables and twisted pairs) to very high-speed optical fiber-based networks. With the advent of the wireless communications era in the late 1980s, the IEEE 802 standard began to develop standards for wireless networks [28]. This development evolved into a number of distinctive wireless standards, covering WMANs to WLANs to WPANs. The main characteristics of these families of standards will be presented and discussed in this section. The IEEE 802 wireless standards organization is an open forum, where the activities are chiefly carried out by industry and academia representatives. While technical discussions are the main motivation of the meetings, the final decisions are also colored by business and political issues. The scale of the wireless developments and standardization in IEEE 802 can be regarded as global, though its center of mass is clearly North America. Indeed, North America contributes with more than 70% of the overall efforts, whereas Asia and Europe are the most active regions in the rest of the world.

In terms of coverage, the IEEE 802 wireless standards range from a few meters to several kilometers; they support low to moderate mobility. As with any standard, one of the principal advantages of having a set of worldwide common specifications is the cost reductions resulting from the economy of scale, which will help to develop the markets. Of the various standards currently being developed within IEEE task groups, the IEEE 802.11 [22] (Wireless Local Area Network, WLAN) and IEEE 802.16 [21] (Wireless Metropolitan Area Network, WMAN) are perhaps the ones attracting the highest share of participation. Compared with 2/3G cellular systems that support communications in a high-speed mobile environment while being able to provide only low/medium spectral efficiency and link capacity, these systems can provide higher efficiency and capacity in fixed/nomadic environments. Recently, the task groups are trying to evolve the previous versions to further improve link capacity and/or to provide more mobility. These activities aim to position these standards as a pre-4G solution. The most 4G relevant IEEE 802 wireless standards are briefly presented next.

IEEE 802.11

The IEEE 802.11 standard [22] can be considered as the WLAN standard par excellence, providing seamless wireless access within the home, office building, and public spaces. This standard defines *ad hoc* and *infrastructure*-based network configurations that exploit the unlicensed Industrial, Scientific and Medical (ISM) bands. Several physical layers are defined in this standard. The plain 802.11 initially defined two radio and one infrared (IR) physical layers. The radio-based link uses the 2.4-GHz frequency band and is based either on direct sequence or frequency hopping spread spectrum techniques. A maximum data throughput of 2 Mbps is achievable in each case. Several evolution versions were later developed supporting higher data rates e.g., 802.11b (maximum data rate of 11 Mbps), 802.11a and 803.11g (up to 54 Mbps) and IEEE 802.11n (up to several hundred megabytes per second).

IEEE 802.15

The IEEE 802.15 standard [23] is a short-range wireless communication standard supporting WPANs. This standard advocates the use of low-power, low-cost, and small-size networks to provide communications within the operating space surrounding a user. Originally this standard followed closely the Bluetooth [24] specifications. This low-rate standard was initially conceived to provide connectivity to a variety of handheld and portable devices like PDAs and notebook computers, as well as peripherals and other electronic equipment. A high-speed WPAN was then developed within the framework of IEEE 802.15.3 providing up to 55 Mbps of data throughput in a range of less than 10m. These networks operate in the unlicensed 2.4-GHz frequency band. The

IEEE 802.15.4 specifies a simple, low-cost, very low-power WPAN and low transfer speeds (20 to 250 Kbps) aimed at wireless personal area and device-to-device networking. Basically, this standard provides wireless connectivity to sensors and actuators, supporting in general tasks of monitoring and control. ZigBee [25] is a parallel standard, sharing the physical layer with 802.15.4 and providing enhancements in higher layers.

IEEE 802.16

The IEEE 802.16 standard [21, 29] specifies a broadband wireless access (BWA) specifically for metropolitan area networks (MANs). Initially the IEEE 802.16 standard addressed the fixed wireless markets (i.e., fixed BWA, also known as wireless local loop (WLL)), aiming to replace with a wireless access the existing wired broadband connections to the core network (e.g., cable modems, DSL, T1/E1 wired access). More recently, the IEEE 802.16 standard has also focused on mobile BWA, concentrating mainly on low-to-moderate mobility subscribers requiring high-speed connections. The 802.16 standard encompasses both licensed and license-exempted bands and exploits single and multicarrier techniques. A large number of industrial partners (more than 200 in 2005) who support its current version (IEEE802.16e) have formed a consortium, the WiMAX (for Wireless Microwave Access) [26], aiming to promote and certify compatibility and interoperability of products complying with the 802.16 specifications. In Korea, the WiBro system is due to be deployed in 2005 and ready to provide services from mid-2006. Based on the IEEE 802.16e technology, WiBro will provide up to 1-Mbps-connectivity mobile terminals, with velocities of up to 60 km/h.

IEEE 802.20

The IEEE 802.20 standard [27] is a mobile broadband wireless access (MBWA) standard enabling worldwide deployment of affordable, ubiquitous, always-on, and interoperable multivendor mobile broadband wireless access networks that meet the needs of business and residential end-user markets. The low-latency air interface is designed specifically for packet transfer, and it is optimized for IP-based services, supporting peak data rates exceeding 1 Mbps. The 802.20 standard operates in the licensed bands below 3.5 GHz. Wide coverage, with cell sizes of 15 km or larger and high mobility are also supported.

5.4   Discussions and Conclusion

This chapter has reviewed the major 4G research and development activities on a worldwide scale. Asia appears to be perhaps slightly more active, although Europe and America are also heavily involved. As already discussed in Chapter

2, 4G defines several operating scenarios (e.g., fixed, nomadic, mobile) with associated communication capabilities (e.g., low-, mid-, and high-data rates). The research and development work currently in progress covers all these areas. However, the efforts are notoriously fragmented, both industry and technology-wise. Industry is approaching 4G by following several paths, and this will result in several complementary and competing technologies. WWRF is certainly the largest consensus building force, and it has already demonstrated that there exists a basic understanding of 4G among industry and academia across the world. WWRF has identified promising operating scenarios as well as suitable technical solutions with potential to become the key building blocks of 4G. In principle, these visions are shared by the majority of the mobile communications industry. Other important (regional) research initiatives (e.g., in China, Japan, and Korea) primarily attempt to fulfill the needs of local markets, though they may also have ambitions to reach other regions. Europe appears to be well organized in terms of a few large-scale research projects and a large number of smaller complementary projects. In general all the aforementioned research initiatives consider wide access (e.g., cellular) and local access as the main constituents of 4G, with different technical solutions in each case. On the cellular side, there seems to be an understanding that conventional architectures may not be enough to guarantee wide coverage and high-data rates. New network configurations are being studied by these initiatives, particularly multihop cellular systems (e.g., typically two-hop or relaying systems) and to a lesser extent, distributed radio approaches. As for the local area, there are several concepts being considered. Sometimes these can be regarded as complementary systems for short-range communications (e.g., PAN, WLAN, MAN). Different competing technologies are being studied (e.g., single and multicarrier, UWB, optical). Most initiatives identify multiantenna systems as one of the fundamental enabling technologies for 4G, in both cellular and local access approaches. This aspect is also seen in the development front, where most, if not all systems being designed or already demonstrated, exploit multiples antennas.

Several challenges can be mentioned. Assuming that several technologies dominate on the 4G horizon, the integration of several networks into one monolithic network that assures a transparent and seamless experience to the user is an intricate task. In many cases the initiatives concentrate only in isolated technical areas rather than the integration of heterogeneous technologies. Also, the practical scenarios and future services already identified and being used as the blueprint for the technical development have not been deeply explored nor validated. Furthermore, their applicability cannot be assumed to be universal. This may lead to risky situations as the technical solutions investigated by ongoing research and development efforts try to match the demands raised from several assumed scenarios and services. Another point of conflict is the fact that the supported technologies and schedules for adoption and launching of the technology are somewhat different in different regions of the world, with Asia pushing for launching 4G around 2010 and Europe considering 2015 or later.

Since 4G is expected to have a highly integrative role, embracing virtually all possible combinations of scenarios and capabilities, the early signs of fragmentation should not necessarily be seen as discouraging, nor necessarily having any negative connotation. Rather, the fact that recent history is repeating (e.g., global technology breakup in 2G, 3G) is once again a clear indication that technology is just one part of the wireless communications picture. To a great extent, political aspects and economic interests are likely to have a decisive impact on the technical solutions being adopted for future 4G wireless communication systems. Even though a universal solution applicable in all scenarios is proposed, the chances for such a concept to become the *de facto* single 4G solution seem to be small, mainly because of the several and nonaligned interests governing the wireless communications arena. The ITU-R Framework Recommendation M.1645 defines a broad range of operating scenarios and capabilities (see Figure 2.12 in Chapter 2) for these systems. Given the supported variety of scenarios and capabilities, and taking into account the number and nature of the players implicated in 4G, it is likely that the number of proponent technologies and solutions will be initially large, as can already be seen. Attention should be paid to further harmonizing this prestandardization work, aiming to achieve wide consensus on a few promising constituent technologies.

References

[1]   Bauer, F., et al., "Synthesis Report on Worldwide Research on 4G systems," IST-MATRICE deliverable D7.1, September 2003, http://ist-matrice.org.

[2]   Wireless World Research Forum, http://www.wireless-world-research.org.

[3]   Wireless World Research Forum, "Book of Visions," http://www.wireless-world-research.org/save/BookofVisions2000.html.

[4]   Mobile IT Forum, http://www.mitf.org/index_e.html.

[5]   Mobile IT Forum, "Flying Carpet Report," http://www.mitf.org/public_e/archives/index.html.

[6]   FuTURE Project, http://future.863.org.cn.

[7]   You, X.-H., and Z. Ping, "FuTURE Project: Toward Beyond 3G," Ninth WWRF Meeting, Zurich, July 2003.

[8]   Next Generation Mobile Communications Forum, http://www.ngmcforum.org.

[9]   MAGNET Project, http://www.ist-magnet.org.

[10] WINNER Project, https://www.ist-winner.org.

[11] Wireless World Initiative, http://www.wireless-world-initiative.org.

[12] Ambient Networks Project, http://www.ambient-networks.org.

[13] End-to-End Reconfigurability Project, http://www.e2r.motlabs.com.

[14] The eMobility Mobile and Wireless Communications and Technology Platform, http://www.emobility.eu.org.

[15] 4G Mobile Forum, http://delson.org/4gmobile.

[16] DARPA Project, http://www.darpa.mil/ato/programs/xg/overview.htm.

[17] Oxygen Project, http://oxygen.lcs.mit.edu.

[18] FUTURA Project, http://www.cwc.oulu.fi/projects/Futura.

[19] Virtual Centre of Excellence in Mobile and Personal Communications, http://www.mobilevce.com.

[20] Sorger, U., I. De Broek, and M. Schnell, "Interleaved FDMA-A New Spread-Spectrum Multiple-Access Scheme," Proceedings of International Conference on Communications, ICC'1998, Vol. 2, June 1998, pp. 1013-1017.

[21] IEEE 802.16, http://www.ieee802.org/16/.

[22] IEEE 802.11, http://www.ieee802.org/11/.

[23] IEEE 802.15, http://www.ieee802.org/15/.

[24] Bluetooth, http://www.bluetooth.com.

[25] The ZigBee Alliance, http://www.zigbee.org.

[26] The WiMAX Forum, http://www.wimaxforum.org.

[27] IEEE 802.20, http://www.ieee802.org/20/.

[28] Marks, R. B., I. C. Gifford, and B. O'Hara, "Standards in IEEE 802: Unleash the Wireless Internet," IEEE Microwave Magazine, Vol. 2, No. 2, June 2001, pp. 46-56.

[29] Eklund, R., et al., IEEE Standard 802.16, "A Technical Overview of the Wireless MAN Air Interface for Broadband Wireless Access."

6

4G Terminals

6.1 Introduction

This chapter explores 4G terminals from the technology and application view-points, in an attempt to identify possible trends, promising technical solutions, and challenges for designers of future portable communications equipment. The critical success factors for future 4G terminals are *service convergence, user-centric interface,* and *portable intelligence,* as shown in Figure 6.1. Service convergence is important not only in the sense of a terminal's ability to support multiples services, some of then concurrently, but also from the standpoint that many services

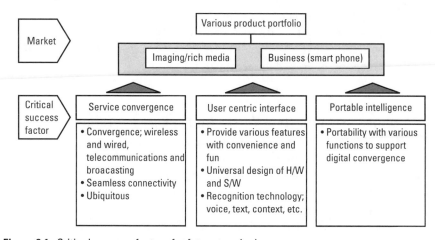

Figure 6.1 Critical success factors for future terminals.

213

from different providers will have to be compatible with a large variety of terminals with different capabilities. We already experience the convergence of various multimedia and interactive services, such as mobile Internet services, multimedia messaging services, and digital multimedia broadcasting (DMB). Concrete examples include: sending short messaging services (SMS) to broadcasters while watching TV on a mobile terminal; voting or any other user interaction on the move during a broadcasted program; and interactive Internet shopping for an object appearing in a MPEG4 movie. As convergence of services and supporting technologies becomes more and more important, the seamless connectivity among these services and technologies becomes crucial. Moreover, as mobile terminals become more complex, it is expected that convenient features aimed at enhancing their usability will emerge-for example, advance recognition technologies and user friendly interfaces. Like the whole 4G concept, terminals must also be designed around the user. A user-centric interface is the first step to fulfill this requirement. Advanced multifunction terminals (and their counterpart services) will not appeal to users if manufacturers (and service developers) ignore friendly and intuitive user interfaces. These requirements must be added to mobile terminals while maintaining portability, which is an inherent characteristic of a mobile terminal. In terms of markets, there is no single terminal capable of fulfilling the need of all users. Indeed, different user segments have different needs, and hence, more than ever can one expect diversity in mobile terminals, such as devices intended for entertainment, information access, and business. It follows that various terminals reflecting the above mentioned critical success factors should be developed. For example, general-purpose terminals targeting the horizontal market need to support imaging and rich media for entertainment. Furthermore, the vertical market will be driven by specialized devices such as smart-phones with business applications like Personal Information Management System (PIMS) and office functions, or embedded terminals for logistics.

6.2   Multimode (All-in-One) Versus Single Purpose Terminals

When we look at the evolution of mobile terminals, we notice that two basic design philosophies are widely accepted. One philosophy aims at *multimode terminals*, where a variety of different functionalities are supported. The other approach targets *single-mode terminals*, where basic telephony functions are supported. Most people feel that users should be able to experience various content, such as music and video, beyond simple telephony through one terminal with multifunctionality. Such characteristics can be seen as fulfilling the changing and somewhat difficult-to-predict needs of an important user segment, the younger generation. However, there does seem to be another important market with less inclination to use such an array of functions; this group is rather satisfied with simple devices equipped only with a minimum amount of essential functions. Thus, one may argue that simpler terminals should be still in the market even if they have as few multimedia functions as possible.

Market trends and fierce competition among manufacturers are manifested as an ever-increasing tendency to incorporate more functionality within terminals. Even though some users would be satisfied with simple mobile terminals, they may end up purchasing devices with imaging and/or music reproduction capabilities incorporated (e.g., built-in camera, MP3 player) simply because those multifunctional (or multimedia) phones are being offered. Note that manufacturers are directly or indirectly starting to support multimode terminals, as this approach could become a very effective way of differentiating from their competitors. In fact, in terms of business market, as the mobile market grows rapidly and the distinction between vendors in technology fade, vendors will focus on the development of multifunctional terminals, possibly combined with aggressive branding, as an effective method for distinguishing from other products.

As time goes by, mobile subscribers will be able to communicate at any time by voice, video, or text messaging, and will be able to obtain various kinds of information and enjoy rich multimedia content. Vast amounts of rich content are already available in the domain of fixed Internet, and these can be also accessed from the mobile network. Dedicated content that matches the needs of users on the move will increasingly be added. The availability of multimedia content and their easy access will foster the development of multifunctional terminals.

The introduction of multifunctional mobile terminals has been accompanied by a rapid evolution in support for multimedia. The real driver originates from audio and imaging applications made popular through other devices [e.g., MP3 players, portable multimedia player (PMP), and Garmin's GPS receiver] and through advancements in technology that allow for cost-effective miniaturization. It comes as no surprise that the mobile phone is the most adequate device for multimedia on the go.

The new generation of mobile phones should be designed taking into account user needs, market trends, and technology evolution. Market trends are difficult to predict, because users, operators, terminal manufacturers, and the state of local and global economies, as well as many other factors, have a big impact on how the markets will develop. Since the user is the final consumer in the terminal market chain, it is precisely the user who ultimately the manufacturer should listen to when it comes to designing future terminals. Diverse enabling technologies such as semiconductor technology for ASICs, memory and processors, displays, built-in cameras, batteries, algorithms, and coding formats, are just means to implement the terminals, and they remain unknown to the user.

Large and bright color displays have become a strong selling point for mobile phones, and emerging features and services make excellent use of them, for instance, graphical user interfaces (GUI), imaging, browsing, and gaming. The display is one of the most expensive components in a phone. Display technology is rapidly evolving and the QVGA displays (320 × 240-pixel resolution) introduced in phones in 2003 will become commonplace in 2005 and 2006.

Also, in a few years built-in cameras supporting still and video imaging will became a must-have feature in mobile phones. As costs continue to fall, many of the standard features associated with original digital still cameras will show up in the mainstream mobile phones-for example, several megapixel sensors, flash, auto-focus, and optical zoom. The added value here, as compared with conventional digital cameras, is the possibility to transfer images straight away by the same device.

In the audio domain, audio codecs such as MP3, advanced audio coding (AAC), RealAudio (RA), and Windows Media Audio (WMA) will be played in the terminal. Most of the current generation phones support stereo headsets, and recently phones with stereo speakers were introduced.

A new voice service, Push-to-Talk over Cellular (PoC) and Push-to-Multimedia, is based on IP transmission of the voice signal without using a circuit-switched connection. A PoC resembles a walkie-talkie type of service that connects multiple users just by pushing a button. As this trend continues, we may see services like full-duplex voice over IP (VoIP) and multiple voice sessions over IP in the market.

Mobile phones in the market appear to follow the trend that more and more features and functionalities will be added-phones, digital cameras, music and video players, digital TV, gaming consoles, positioning, messaging clients, and PDAs. The mobile phone hardware and software will thus become increasingly more complex. Obviously, to be successful in future mobile markets, manufacturers need solid and flexible hardware and software platform architectures.

6.3   Future Terminals and Technology

In the previous section two approaches to 4G terminals were discussed: single-mode versus multimode (multiple functions) terminals. At this point it appears impossible to predict if any of these two approaches will emerge as the predominant one. Based on the present situation one could argue that both approaches could share the market. However, it should be kept in mind that one of the driving forces of future wireless communication systems is their capability to effectively handle multimedia type of information and the related services and applications, and hence, flexible, multipurpose devices seem to be the most promising candidates to become the mainstream terminal of the future.

Terminals will come in many shapes, sizes, and capabilities in the 4G world. *Heterogeneity* may be the key word that best describes the realm of 4G terminals. Some will be inexpensive, simple, and with very basic communications capabilities; others will be advanced and expensive, boasting truly high data throughputs. For illustrative purposes, Figure 6.2 shows the broad range of terminals for 4G and their principal characteristics [3]. As can be observed, in the context of 4G, terminals refer not only to the conventional mobile (or portable) phones but also to an array of very different and capable devices. Figure 6.2 includes nonconventional devices that can be thought of as terminals, just by the fact that a communication enabling transceiver is integrated to those devices, and also because other related functions may also be incorporated (IP address, display, etc.)

Demands on flexibility, adaptability, user friendliness, operational time, and radiation levels are driving the development [2]. In this sense, 4G terminals will require more developed features than the existing terminals, including highperformance, multifunction, low-power consumption, internetworking capability, wide bandwidth, miniaturization, and others, as summarized in Figure 6.3.

Next, a list of the expected basic requirements of hardware and software technologies for the 4G terminals is presented.

Hardware

- Multimedia;
- Modem, processor: enhancing processing power, embedding application processor;
- Display, camera, storage, battery: all requiring high-quality and high-performance because of the rich-media services;
- Flexibility to various environments;

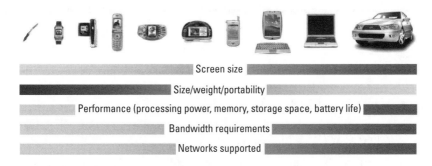

Figure 6.2 Broad range of heterogeneous terminals for 4G. (From: [3].)

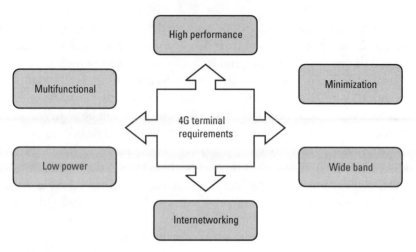

Figure 6.3 Basic requirements for 4G terminals.

- RF, antenna: support of multiband and multimode, support of advanced technologies (e.g., SDR, MIMO, smart antennas);
- Modem: System-on-chip (SOC) (i.e., modem and RF integrated onto one chip);
- Multiple connectivity: supporting several communication standards (e.g., W-CDMA, CDMA2000, Wireless LAN, Bluetooth, WiBro);
- Active development of the alternative technologies;
- Display, keypad: maximize portability and performance (e.g., flexible display, virtual keypad, virtual display);
- Storage, battery: overcome the limitations of current technology (e.g., small HDD, MRAM, FRAM, fuel cell);
- Convergence;
- Camera: auto-focusing, several megapixel resolution, and optical zoom;
- Various technologies and services converged into the mobile terminal, such as media, biometric, health, leisure, and fashion.

Software

- General purpose, flexible;
- Connectivity protocol: seamless connectivity among W-CDMA, CDMA, WLAN, Bluetooth, and WiBro;
- Middleware: wider acceptance of middleware platform, such as Java and Brew;

- Operating system (OS), codec: open OS (Symbian, WinCE, Palm) and general purpose Codec (H.264);
- Standardization is very competitive;
- Middleware: Java versus Brew;
- OS, codec: Symbian versus MS, H.264 versus MPEG4, MP3 versus AAC/WMA.

Markets are expected to evolve from network-centric environments towards terminal-centric environments. In other words, so far the critical issues have been how to provide and deploy a high-speed network with QoS support to the end user. These considerations are solved by broadband network technologies like 3G and WiBro. Once the problem of providing network connectivity in several environments is solved, the next challenge is to bring various broadband multimedia services and applications into the mobile terminal. The service provided by a network will often be determined by the capabilities of the terminal being carried by the user. Basic and simple terminals will support basic and simple services and vice versa. The same services but with upgraded capabilities could be supported by advanced terminals (e.g., higher resolution due to better and bigger displays and higher computational power). As pointed out, terminals could also be used for many new purposes, including functions like MP3 player, radio receiver, TV receiver, camera, remote control, data storage, and positioning. Also, various software platforms for mobile terminals are already in use or expected to be used in the future, such as JAVA, Symbian, Microsoft Windows, and Linux. The terminals will have similar applications to those found in PCs, but more and more often third parties will provide applications that use the terminal as a platform for many new purposes. This evolution requires that the programming languages used in the terminals can use most of the functionalities that the terminal possesses (i.e., more interfaces). Another challenging issue is the interoperability of applications and services in terminals that have different capabilities (display resolution, processing power), operating systems, and software platforms. It is likely that there will be many competing operating systems in the terminals at first, but eventually only a few will have a leading role in the markets. Figure 6.4 [1] presents an overview of hardware evolution and technology development for future terminals.

The user interface evolution consists of displays, multimodal interfaces, and standard applications, such as user positioning (e.g., GPS), camera, and biometric identification. Multimodal interfaces for man-machine communication can be regarded as one of the key technologies, as they determine the appearance and interaction of a device or application and will be optional to keyboard and mouse. Pen, speech, and gesture-based operation/interfaces will become increasingly common in mobile consumer terminals.

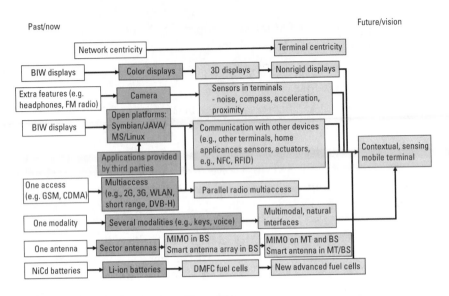

Figure 6.4 Mobile terminal technology. (From: [1].)

At present, terminals are only capable of accessing one network at a time and a terminal is capable of switching a radio network among some standards, like GSM, UMTS, CDMA, WLAN, and WiMAX. Parallel radio or multiple simultaneous access technologies will allow access to different radio networks at the same time. For instance, one can be connected to the cellular network engaged in a voce service while simultaneously retrieving information from a WLAN. Antenna evolution promises increased throughput and more optimal utilization of bandwidth using MIMO-capable base stations and mobile terminals. Terminals will be able to communicate with a large number of other devices using different techniques (e.g., radio and optical interfaces) [4]. Many types of sensors will be embedded in future devices to allow control of applications and to sense the current user context.

It is expected that future terminals will have higher power consumption due to the new high-speed multimedia applications, the multiple radio interfaces that could be used simultaneously, the increased processing power required by broadband multimedia-optimized terminals, and the use of multiple antennas and large displays. To get an insight of the power requirements of mobile phones, a typical 2G terminal needs typically about 10 MIPS (Million Instructions Per Second), while a 3G terminal may require a few 100 MIPS. It is expected that 4G terminals may require a computational power of one or two orders of magnitudes higher that their 3G counterparts. Unfortunately, higher MIPS means higher power drain. In terms of power/volume, today's battery technology is not sufficient to power the devices foreseen for the future. This is an issue that could become critical and eventually it could hinder the development of 4G terminals. A very advanced high-performance terminal will be far less appealing to users if its operational time is very short or if it has to be powered by bulky and heavy batteries, despite all of its capabilities. Low-temperature fuel cells, such as the direct methanol fuel cell (DMFC), appear to be a promising solution for replacing conventional rechargeable batteries and solving this challenge, though technical developments in this area move at a much slower pace than in other technology fields related to 4G terminals. Even if the power issue is solved, however, another critical issue is the extensive heat production and necessary cooling for handheld terminals [3]. Battery technology is a fundamental area that could slow down the development of truly advanced terminals.

6.3.1 Hardware Technology Roadmap

A hardware technology roadmap is shown in Figure 6.5. Several key hardware-related components and technologies are discussed next.

Modem

As the need for processing power is increased, the competition between integrated single-chip architecture and functionally divided multichip architecture will become more pronounced. Single-chip architecture has the advantages of requiring less size and in some situations lower power consumption. On the other hand, multichip architecture is powerful in processing multimedia applications. Some studies show that by distributing processing power, a significant reduction in power consumption can be attained. There is no clear rule that shows which approach is the best, and in practice, particular cases should be evaluated individually. Also, the need of fixed (e.g., ASICs) or flexible (e.g., reconfigurable) structures has to be assessed carefully when designing the platforms for implementing the modem. Cost and space limitations also need to be taken into account when considering single and multichip architectures.

Radio Technology

Terminal architecture will be designed by means of a modular structure integrating various functions and radio access technologies in small-size mobile terminals. New advanced technology shall be developed to support multiband/multimode like W-CDMA, EV-DV, HSDPA, DMB, WLAN, WiBro, and so forth. SDR and multiantenna technologies will become the core technologies, offering flexibility to the changing radio environments and high performance.

In particular, SDR enables easier expansion of base station facilities, flexible reconfiguration of air interfaces, and terminals to support multiple standards. The flexibility offered by this technology is paramount to make possible

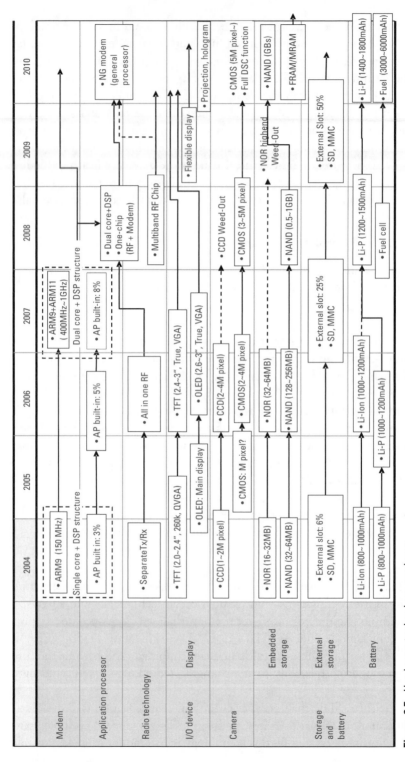

Figure 6.5 Hardware technology roadmap.

truly seamless networks and to support multiple communication schemes. It also provides an effective means for expanding functions for multimode operations, as terminals can be reconfigured in the field. The changes of terminal features can be carried out even without users being aware of them through software download [5]. In addition, multiple antennas at both ends of the wireless channel can be used to increase the data rate via spatial multiplexing and to improve the link quality through diversity. The integration of multiple antennas into mobiles will result in many very attractive features.

I/O Devices

The display will be an important part of a 4G terminal in providing multimedia services in the ubiquitous environment and in offering a realistic and vivid experience exploiting a high-definition imaging capable device. Various types of displays like flexible displays, micro displays, and hologram-based devices will also be developed for various service environments. Portability is a key factor for a mobile terminal and current technologies like multiple key presses or touch screen are not necessarily comfortable and easy-to-use input devices. In the future, more convenient and easy-to-use keypads will be required and, external/virtual keyboard as well as intelligent text-input will be spotlighted. Image sensors, microphones, and other sensors, through advanced gesture, voice, and movement recognition, will also be considered as input devices.

Storage and Battery

The multimedia capabilities in phones and camera phones are speeding up the development of storage, such as flash memory, micro-hard-disk drives (HDD), and external memory cards. In the near future, solid-state devices like NOR flash memories will be in the mainstream with low cost and large capacities. Indeed, capacity on the order of several gigabytes will be built-in onboard the mobile terminal to support rich-media contents. Battery technology is one of the necessary key technologies for realizing such appealing mobile communications terminals. The technical challenges in this area include the enhancement of battery capacity, power efficiency, operational hours, and lifetime (number of charging times), and a reduction of size and price in order to improve their portability and ease of use [5].

6.3.2 Software Technology Roadmap and Issues

Complementing the hardware roadmap shown in Figure 6.5, Figure 6.6 shows the software technology roadmap for future terminals. The need for a high-speed and broadband network is already recognized, and 3G evolution, wireless broadband (e.g., WiBro), and 4G communication consider such requirements in their initial assumptions. Wireless broadband will make content

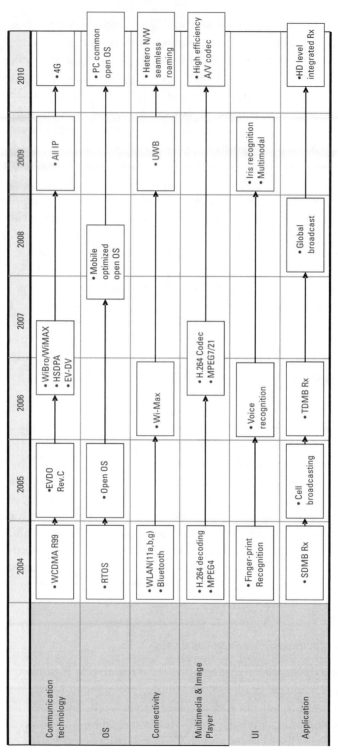

Figure 6.6 Software technology roadmap.

and services rich and encourage the development of multifunction and high performance OS. Besides, various services exploiting broadcast and user location will converge into mobile terminals, together with easy-to-use and comfortable user interfaces.

Communication Technology

The 4G mobile systems are expected to offer higher transmission capabilities in different phases building upon the evolution from the 2G and 3G systems and providing users with an optimal access depending on their locations through linkage of multiple systems. Transmission speed of several megabytes per second per user will be also required in cellular environments due to the projected proliferation of wired Internet and multimedia services among the public. In addition, in hotspots a speed of about several tens of megabytes per second-about 10 times faster than the speed available today-is expected to be required. These requirements can be seen as the driving force for 3G evolution like HSDPA and EV-DV and wireless broadband services like WiBro and WiMAX. Besides, short-range networking technologies like Bluetooth, WLAN, and optical wireless communication systems (e.g., IrDA) will play a significant role in communicating among various devices (including mobile terminals) and will accelerate the building of the ubiquitous world. In this respect, seamless networking, which is a technique to connect packet-based backbone networks with other radio access systems or networks, is essential for users to use multimedia services through various networking environment simultaneously.

Operating Systems

An embedded real-time OS is generally used in a mobile terminal. However, the advance of multimedia content and multipurpose terminals has stimulated the development of open OS and mobile platforms. The OS and platforms with a user-centric interface and PC compatibility will take competitive advantages in a hot rivalry. Figure 6.7 shows the development trends of terminal platforms and operating systems.

User Interface (UI) Technologies

The emergence of high-speed and broadband networks (such as 3G, WLAN, and WiBro) makes possible the transfer of rich content to the terminals. Moreover, the rapid development of computing technology, such as CPU and DSP, and memory enable mobile terminals to function like personal computers. These changes in the mobile environment will create some pressure to develop technologies that allow a friendly use of complex terminals, including the access of mobile services and applications. UI technologies like signal processing, which enables the use of virtual keyboards and voice and gesture recognition,

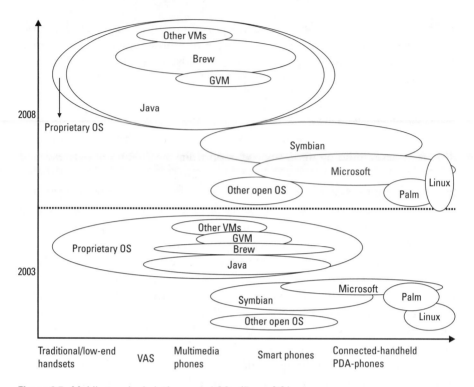

Figure 6.7 Mobile terminal platforms and OSs. (From: [6].)

will be indispensable in eliminating the limitations of mobile terminals like small-size screens and keypads.

Multimedia

In the area of multimedia, the following are some of the most important requirements for reproducing content in superior quality: high-resolution display, high-quality video distribution, high-quality video play, a virtual reality capable of providing a sense of presence, and support of high-fidelity audio. Also required will be functions to change the audio/video (A/V) quality, network conditions, or terminal capabilities (voice/video quality selection). Multimedia has already become a commonplace service in PC environments, but to enable equivalent processing in future mobile platforms, a tremendous leap must be achieved in lowering power consumption and enhancing video processing techniques. To provide QoS, techniques that guarantee users' quality in real time must be developed [5]. Currently, H.264 and MPEG4 are the main multimedia codecs that are optimized for mobile terminals, and highly efficient and high-performance A/V codecs will be developed as the processing power of CPUs and OSs is increased.

Applications

As the digital convergence is accelerated, new services are expected to emerge in several fields, such as location, telemetric, broadcasting, and high-quality multimedia services. Wireless communication will be the most effective method for distributing information and multimedia content to numerous users when broadband access and high speed are available. Broadcast and multicast services enable multiple users to share common information. In addition, they can be used to support new applications that could further stimulate the spread of mobile communications services, including the delivery of electronic papers or news at stations, airports, and hotels, the efficient provision of character information or still/moving images to a number of users (e.g., at business shows), and the delivery of on-demand video/music distribution [5].

6.4 Key Applications for Future Mobile Terminals

Mobile users have always had a positive attitude towards novel and useful applications and services. There is no doubt of the importance of wireless applications and services in the future. Applications and services will shape the development of future terminals as terminal capabilities determine the type and performance of applications and services that the terminal can support. As mentioned, a distinctive characteristic of 4G is the fact that a whole range of different terminals will be used, ranging from simple voice and messaging only, to very advanced and powerful computer-like terminals relying on advanced input/output technology. Thus, applications and services are factors as critical to the success of the future mobile terminal business as the different discussed technologies. Convergence between broadcast and telecommunication is already in progress, and in the future this convergence will be accelerated rapidly to include several different fields, such as broadcast, telecommunication, and computer. Broadcast, Digital Rights Management, Multimedia Messaging Service, over the air provisioning, presence, PoC, and user plane location are important applications to discuss. This section examines future applications and services from the terminal development point of view.

6.4.1 Broadcast

Mobile broadcast services refer to a broad range of services that make use of the one-to-many communication paradigm. Current developments in terminal technologies and digital broadcast systems have made broadcast services possible to use in the mobile environment. This, in turn, will enable low-cost mobile distribution of multimedia content to users in the service area. Consequently, substantial new business opportunities will open up for content and service providers as well as terminal and system vendors.

Digital broadcast technologies such as DVB-T/H [7], ISDB-T [8], 3GPP/MBMS [9], and DAB [10] will make possible the distribution of any digital content through the broadcast channel, such as audio and video streams, movies, application software, or Web pages, with dramatically lower costs. Indeed, the delivery costs of broadcasting are insensitive to the number of receivers (within the coverage area of a transmitter). Consequently, media content can be delivered to large audiences at a fractional unit cost when compared to conventional mobile delivery over two-way, point-to-point wireless networks [11].

Broadcast and telecom network cooperation refers to the joint usage of these two complementary technologies in order to provide new features that each technology individually cannot provide in a satisfactory manner. Such a joint use of both networks (see Figure 6.8) can improve the capabilities and varieties of services and can make services more affordable.

6.4.2 Digital Rights Management

Digital Rights Management (DRM) defines a set of technologies that provides the means to control the distribution and consumption of the digital media objects, as shown in Figure 6.9. Since 4G is expected to rely heavily on multimedia communications, DRM is an issue that needs be taken into account at the early stages of development. The DRM system is independent of media object formats, operating systems, and runtime environments. Content protected by the DRM can be of a wide variety: games, ring tones, photos, music clips, video clips, streaming media, and so forth.

Before content is delivered, it is packaged to protect it from unauthorized access. A content issuer delivers DRM content, and a rights issuer generates a *rights object*. A *rights object* governs how DRM content may be used. DRM

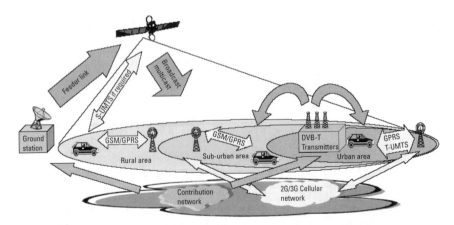

Figure 6.8 Overview of networks. (From: [12].)

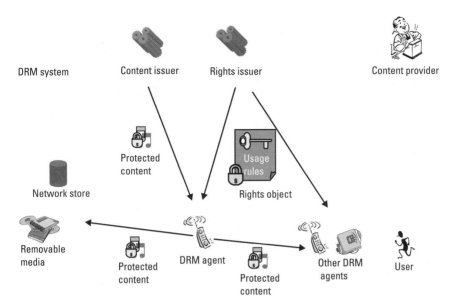

Figure 6.9 OMA DRM functional architecture. (From: [20].)

content and rights objects may be requested separately or together, and they may be delivered separately or at the same time.

DRM content can only be accessed with a valid rights object, and so it can be freely distributed. This enables, for example, super distribution, as users can freely pass DRM content between them. To access the DRM content on a new device, a new rights object has to be requested and delivered to a DRM agent on that device.

Open Mobile Alliance (OMA) DRM [19] is one of the most well-known standards for mobile devices. OMA DRM v1 is for low value content such as ring tones. OMA DRM v2 is aiming for high value content by providing sophisticated security features for secure delivery of a rights object from a rights issuer to a mobile device. Some features of OMA DRM v2 include the following:

- Enhanced security for the acquisition of rights objects;
- Support for unconnected devices (a device without network capability);
- Content sharing by multiple devices;
- Exporting OMA DRM content and rights objects to other DRM systems.

OMA DRM v2 mainly supports the protection of mobile download services, but it will be applied to mobile broadcast services as well.

6.4.3   Multimedia Messaging Service

Multimedia Messaging Service (MMS), as its name implies, provides a communication technology to exchange multimedia messages between users. In contrast to the well-known SMS, which can only contain limited text, MMS capable handsets can send and receive much larger messages combining, in addition to text, various types of content such as still images, audio, or video. An example is shown in Figure 6.10.

Like many existing messaging systems (e.g., e-mail, SMS), MMS is nonreal-time and therefore relies on a store-and-forward paradigm. Interfaces between MMS and other messaging systems have been defined in various standardization bodies (e.g., 3GPP, OMA) with the expectation of seeing these different services interoperate. Another important aspect of MMS is to establish a service capable of interoperating across different networks, while remaining backward compatible with older version MMS terminals as the service evolves.

The following describes some recent MMS features (standardized in OMA [14]):

- Support for high-quality images: To support the introduction of handsets with high-resolution cameras, MMS is able to support resolutions beyond VGA.

- MMS with service providers: In contrast with traditional person-to-person multimedia messaging, much effort has been put into enhancing interactions between subscribers and service providers. A typical example is the MMS postcard service, where it is possible to send a multimedia message to a service that prints and mails it as a normal postcard to the appropriate recipients. Enhanced media types and presentation methods have been standardized to support multimedia communication exchanges with service providers.

"See what I saw in Paris when I went on holiday."

Played or spoken sound

Display of text and picture

Arc De Triomphe

Figure 6.10 Example message with multimedia content. (From: [13].)

- Extension of MMS to external applications: The potential of MMS can be extended to easily support new external applications and third-party services. For example, two MMS subscribers may be able to play chess via a third party service provider. Although the chess game by itself is not an application related to MMS, it can use MMS as a bearer to provide, say, the chess board between the players.

- Templates: Template can be defined as a model or pattern that allows the MMS subscriber to easily create multimedia messages (MMs), as illustrated in Figure 6.11. For example, an MMS postcard service could provide such a template to subscribers so that they can more easily create postcards.

6.4.4   Over-the-Air (OTA) Provisioning

Recent technological advances in the mobile communication industry have led to the introduction of large-scale Internet style, media-type independent, download of media objects. Examples include screen savers, ring tones, and Java MIDlets. Technology for confirmed download is used to deliver digital content such as entertainment and business applications as well as media objects to be rendered to mobile devices. Another important application area for download is to personalize terminals according to a user's preferences and tastes. Some of the specific goals include the following:

- Enable both automated and manual client driven capability negotiation.

- Allow the initial download solution to be extended with new attributes and functionality.

- Create a solution that is quick and easy to deploy in order to shorten the time to market.

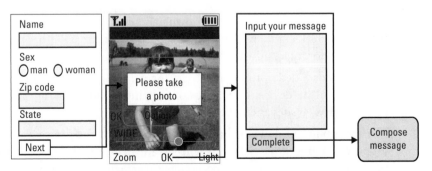

Figure 6.11   Template example. (From: [15].)

- Enable various payment models to support the launch of e-commerce concepts.
- Avoid fragmentation of content space. Content for mobile devices with different capabilities can be published using a consistent concept.
- Create commonality among the download process for all types of media (e.g., games, audio, and pictures).

For the content provider and operator, the main benefits of OTA provisioning are as follows:

- A description of the downloadable media objects can be published on the presentation server in a format independent of the presentation language (e.g., XHTML, WML, HTML) supported by the devices downloading the media objects. The content provider creating the media objects needs only to publish the download descriptors on its Web site; the front-end of the download service can be provided by someone else, such as a special download Web site with media objects from many different content providers.
- The download is confirmed, with the status report posted to the network after the download. It can be used for monitoring the quality of the service and as the basis for billing.
- The user can be directed to a Web location provided in the download descriptor.

For the end user, the main benefits of OTA provisioning are as follows:

- The end user is given a chance to confirm the download based on information in the download descriptor and the device's capabilities; "How much room do I have left for other objects after having downloaded this one?"
- The media object metadata in the download descriptor can be stored in the device and made available in the user interface as "properties" (name, vendor, type, info-URL) of the media object.
- The download descriptor is the basis for a familiar "download user interface" in the device, a sequence of user interactions that are independent of Web sites and networks.
- The probability of downloading a media object that the device cannot support is reduced as capability checks are made in the device just before the download starts. The user is prevented from downloading contents that do not work well on his device.

Last but not least, the concept of a descriptor containing metadata about media objects and Web resources is extensible to use cases other than download.

6.4.5 Presence

The presence information describes the characteristics and the current status of the target user and the user's service(s) and device(s). The presence service collects such information from various sources and then disseminates it to the entity(s) that has requested the information.

As shown in Figure 6.12, the presence service is composed of the following:

- *Presentity*, a logical entity referring to the subject of the presence information. It includes a person, a group of people, a nonhuman entity, or even an abstract entity such as service itself.

- *Presence source*, which contains the information on the presentity and publishes the presence information on behalf of the presentity.

- *Presence server*, which collects the presence information from the presence sources, processes the collected presence information to resolve a consistent view on the presentity, and notifies the watcher(s) such presence information.

- *Watcher*, which receives from the presence server the presence information on the presentity.

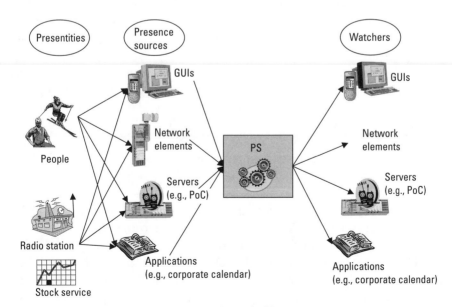

Figure 6.12 Presence service components. (From: [16].)

The presence service enables an entity to recognize its surrounding service environments, allowing the adaptive service behavior. For example, a user wants to reach his friend and has both telephone call and instant messaging service available. By virtue of the presence information on his friend, he finds out that his friend is reachable with the instant messaging and uses it to communicate with him.

Such a service environment-cognizant system is envisaged not only for enhancing existing services but also for expediting the provisioning of richer information and the advent of new value-added services.

6.4.6   PTT over Cellular

PoC, a newly defined terminology in OMA, is a mobile application service that adopts unidirectional communications. It allows mobile users to be engaged in an immediate communication with other users in a one-to-one call or a one-to-many call. As known by its name, one of the characteristic features of PoC service is its half-duplex transmission, and thus it requires floor control mechanism. On this point, it is similar to a walkie-talkie application where a user presses a button to talk with an individual user or broadcast to a group of users. After a speaker sends his voice, the receiving participants either hear the sender's voice without any action on their part (without having to answer the call), or are notified and have to accept the call before hearing the sender's voice. This depends on the answer settings of the receiving users. Other participants can just respond once this initial speech is complete and are allowed to talk to target participants. This feature of half-duplex communication contrasts with a general voice call, which is full duplex, where more than one person can talk simultaneously.

OMA is currently developing many service enablers to bring on various mobile application services. The PoC service is one of the significant and popular application services. Market demands are requiring its urgent development in the near future. This is related to the peculiar features of PoC technology. As a representative application of IP multimedia CN domain (IMS), PoC utilizes many functionalities and infrastructures of that domain. For example, the PoC service enabler takes advantage of the functionality of the presence server to provide ready-to-link service. Also, many types of group calls or conference calls are supported by the provisioning of the group management server in IMS domain. Though the first version of PoC does not include media other than voice stream, its inherent capabilities do not restrict a specific media type as transmission format. This also makes it possible for the PoC service enabler to have a data orientation beyond simple voice service. With its strong coupling to data capabilities, other key features that are also needed for other data messaging services are also supported, like PoC interworking with instant messaging service and conversion of group call type from voice to text conferencing. The first version of PoC in OMA is treating only voice as its communicating media.

From a business point of view, a number of push-to-talk services and their supporting equipment have already emerged in the market. However, to date, these services and the products are all proprietary in nature. This can bring on a market fragmentation and too much diversity in the relevant industry. For global interoperability, a standardized specification is required that defines service in detail. As a final certification process, OMA mandates interoperability tests among different vendors' equipment.

As the first stage of a two-phases approach, the first release of the PoC service enabler is in its completion period. The currently developed PoC service enabler, PoC Rel-1, supports the following capabilities. (1) There are three types of group call including ad hoc group, prearranged group, and chat group, each of which has typical feature of dialing procedure. (2) There is support for incoming session barring and instant personal alert barring. Those barring capabilities would enable the PoC users to prevent unwanted PoC invitation and from receiving unwilling invitation messages. (3) PoC Rel-1 also supports preestablished sessions and simultaneous sessions. Preestablished session is newly introduced to allow early session setup by exchanging media parameters such as codec, IP address, UDP port number, and talk burst control protocol. This preliminary session can provide fast call setup using specific SIP method and RTCP APP message for both originating side and terminating side. Simultaneous session allows participating multiple sessions at the same time, but allows only one media transmitted to the PoC user. This feature will be useful when a participant is joining multiple sessions and wants to move continuously to get information. (4) As procedural features, automatic answer, manual answer, and manual answer override are also provided. Figure 6.13 shows the perspective view of the PoC service for one-to-one call and group call.

Many service aspects are required to implement full functionality of PoC. Figure 6.14 illustrates the architectural block diagram of PoC Rel-1.

6.4.7 User Plane Location

Mobile location services based on the location of mobile devices are becoming increasingly widespread. User plane location employs user plane data bearers (e.g., IP) for transferring location assistance information such as GPS assistance and for carrying positioning technology related protocols between a mobile terminal and the network. User plane location is considered a cost-effective way of transferring location information required for computing the target mobile's location. To offer a location service to a client, considerable signaling and position information is transferred between actors. Currently, assisted-GPS (A-GPS) provides more accurate positioning of a targeted mobile than other available

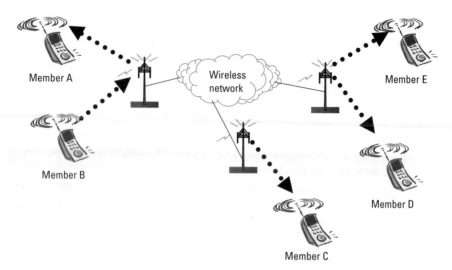

Figure 6.13 Example of a PoC one-to-many group session (voice transmission). (From: [17].)

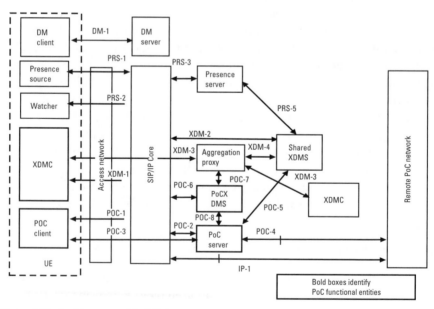

Figure 6.14 Architecture of PoC Rel-1: functional block and interface. (From: [17].)

standardized positioning technologies. However, A-GPS over control plane of
Radio Resource Control (RRC) or Radio Resource LCS Protocol (RRLP)
requires modifications to existing network elements and interfaces (for signaling
procedures between the terminal and the network). Location over user plane

needs only an IP capable network and requires minimum modification to the network, and therefore user plane location is an efficient solution that can be deployed rapidly.

6.5 Challenges

Future 4G terminals could become extremely complex, particularly if the trend of developing multipurpose terminals capable of providing various types of connectivity (e.g., multistandard) and supporting diverse services persists. A number of challenges for the design of such terminals have been identified. Among these challenges, RF integration and the evident mismatch between increased power consumption and the lagging development of battery technologies are considered as crucial. In this section we will discuss these issues and see how they could impact the development of 4G terminals.

RF Integration

One can get a sense of the challenges that designers of future terminals will face by assuming a hypothetical feature-rich terminal boasting typical present-day connectivity and having some receiving functions. In order to provide wide coverage, mobility, and roaming, the terminal would include cellular 2G (e.g., GSM) and 3G (e.g., UMTS) transceivers working in the 900- to 1,800-MHz bands. High-speed local access could be provided by an 802.11 (Wi-Fi) transceiver in the 2.4/5-GHz bands. Basic short-range communication capabilities could be additionally carried out with a Bluetooth transceiver operating at 2.4 GHz. RFID is already being integrated in some handsets and allows very short-range instant connections for several purposes (e.g., payments, identification, information retrieval). RFID operates typically in the 13-MHz band, though low and high-frequency versions are available (100 KHz, 1 GHz, 2 GHz, 5 GHz). In addition, and following the trend of designing terminals with as many features as possible, one can also see the integration of a conventional FM radio receiver (100-MHz band), a terrestrial or satellite broadcast TV receiver (100-MHz to 1-GHz band), and a GPS/Galileo (1.1- to 1.6-GHz band). Though all these services will not be used simultaneously, a few could be switched on in certain applications. Terminal design could become the bottleneck for the development of multistandard multifeatured terminals, as radio frequency interference between different transceivers could have detrimental effects on the performance, in particular taking into account the physical limitations of having receivers and transmitters enclosed in a very small space. Methods for interference management and control should be carefully applied at the design stage. The situation becomes critical if one considers the fact that the transceivers (e.g., for communications) and/or receivers (e.g., radio, TV, GPS) may need to use different antennas. Moreover, in order to attain the expected level of performance, particularly high data throughputs, cellular and local area systems have to resort to the use of multiple antennas. At high frequency bands (e.g., 5 GHz or higher) it is feasible to place several antennas on a small (handheld) terminal-for instance, an antenna array of four or even eight elements. Typically, these antennas have to be distributed to cover virtually all the available main plane (largest surface) of the terminal, in order to maximize antenna element separation. Obviously, future designers will face a very hard task trying to integrate efficiently multiple receiving and transmitting elements into small form factors. This could even be impossible for certain combinations of multistandard and desired form factors. There is another important factor against having implemented several transceivers on one terminal: cost. Even in the case where multiple integration is technologically feasible, the relative high cost associated with having several standards implemented remains one of the central issues to be solved. However, there exists a different approach that has been extensively studied in recent years, and in the view of their proponents it could be, cost- and technology-wise, the most efficient way to go towards making complex terminals a reality. This solution is known as Software Defined Radio. Ideally, and in its broadest sense, SDR allows on-the-fly modifications of the RF front ends, baseband processing, and even the MAC layer of the terminal, aiming to realize a given air interface by reconfiguring the system. The degree of flexibility brought by real-time reconfigurability opens up a new world of possibilities for users, operators, services providers, and terminal manufacturers. Users can establish a connection in any network, allowing simple local and global roaming. Users could also benefit from the low-cost terminals that this technology would eventually bring. As new advanced services are introduced, software modifications can catch up with the new requirements. Hardware and software updates could be easily and wirelessly carried out by the user or by operators. Manufacturers can also take advantage of SDR since large volumes of terminals with identical hardware (and fewer number of components) would have to be produced. Even upgrades or changes in the terminals could be carried out simply. Also operators and service providers could exploit this flexibility to better match their operation and services to user demands.

Mismatch Between Increased Power Consumption and Development of Battery Technology

In addition to the RF challenges described above, there are several fundamental issues that need to be solved to allow the full development and further success of 4G terminals. One of the most critical challenges comes from battery issues, and more precisely, from the slower development of battery technology in comparison to developments in communications technologies. The obvious wireless or mobile condition of a terminal is ultimately dictated by the battery simply because the terminal operates wirelessly as long as its battery will allow it. Several of the most important features of 4G pose fundamental questions to the designers of future terminals, as one could expect that the power consumption of such terminals will dramatically rise. One can already notice this tendency in current 3G terminals and advanced PDAs as multimedia functions and high-speed communications are draining power to such levels that battery life is seriously compromised. Thus, one can easily foresee that 4G terminals will have a much higher consumption. First, the support of broadband high-data rates means that more energy is needed to convey the information bearing bits reliably over the radio channel. High-speed processors are also needed for fast wideband data processing, which has a direct impact on power consumption. In addition, it is generally agreed that multiple antennas will play a fundamental role at the terminal side in order to achieve high data rates and guarantee an acceptable quality of service. This means that the terminals will integrate a number of transceivers, typically two, three, four, or even more. Certainly, more transceivers mean increased power consumption. Multimedia services and any type of transfer of rich content, including broadcast services, inherently consume a lot of power in their use of displays. High-definition video services imply more pixels, improved contrast, and more computational power to process more frames per second, all of which will exhibit increased power consumption. One could then add the previously discussed fact that future terminals are likely to incorporate various functions, with many of them being used simultaneously by the user or application. To this end one should remember that 4G advocates the concept of always being connected, meaning that the battery is constantly being drained.

From the standpoint of 4G terminals, the key desirable characteristic of a battery would be its very high energy density (i.e., measured typically in Watt-hour per kilogram). Other desirable characteristics are long life-cycles, fast charge capabilities, safety, and environmental friendliness (low environmental load or low toxicity). The energy density of typical batteries is as follows: Nickel-Cadmium (Ni-Cd) 50 to 80 Wh/kg, Nickel-Metal Hybrid (Ni-MH) 60 to 120 Wh/kg, Lithium-Ion (Li-Ion) 120 to 180 Wh/kg, Lithium-Ion-Polymer (Li-Polymer) 120 to 250 Wh/kg, and future fuel cells from 500 Wh/kg to more than 1,000 Wh/kg. Present-day terminals mostly use rather high energy density batteries, usually Li-Ion and Li-Polymer based cells. In term of capacity, fuel cells have in theory the potential to fulfill the higher power consumption demand of future 4G terminals. However, and despite huge research efforts, the development of the fuel cell is still far from being mature. Miniaturization and weight and cost reduction are perhaps the most challenging issues that need to be solved before the fuel cell finds its way into future 4G terminals. The development of 4G as a whole will not be slowed down by the battery problem, but conceivably the development of 4G terminals could be slower than expected. This could be particularly true for portable advanced models with very-high-speed capabilities, as they would need to incorporate several technologies prone to increased power consumption. On the other hand, as we have already discussed, 4G terminals will be represented by a heterogeneous array of devices. In some cases like laptops with communication capabilities (another type of 4G terminal), the power drain will not be so critical since battery capacity is significantly higher than in small terminals. It is likely that upon introduction of 4G, small terminals will exhibit moderate capabilities, nevertheless outperforming 3G terminals. As battery efficiency is enhanced-hopefully within a few years from the launch of 4G-manufactures will be able to offer better capabilities. We expect that two-antenna terminals will be typical for the early years of 4G.

6.6 Conclusions

In this chapter we discussed 4G terminals from different perspectives. The actual trend of having multifunction terminals is likely to continue. Even today, voice-only terminals are rapidly being displaced by devices with imaging and high-quality audio reproduction capabilities. 4G supports broadband communications, the main driving force behind multimedia services. Certainly, 4G services can be better exploited in multifunction terminals. However, there is also a place for simple and basic terminals within 4G. In general we can say that a rich and diverse array of terminals will operate in 4G networks. Heterogeneity in terminals means not only a wide range in capabilities and performance, but also in size and shapes. Advanced multifunction terminals represent a challenge for the designers since portability and high performance (e.g., high-speed communications, high computational power) have to be provided. Advances in display technology, availability of larger and cheaper memory devices, miniaturization of components, antenna technologies, and new user friendly man-machine interfaces together have the potential to conjure up in a class of terminals that would allure the user. Hardware and software technology that will find its way in 4G terminals was presented and discussed in this chapter. Several challenges that could slow down the development of advanced 4G terminals were identified, among them the problems of RF integration and low battery efficiency. In addition, another challenge related to terminals is how to adapt and scale services according to the type and capabilities of the terminal requesting a particular service. This is a difficult integration question, in particular taking into account heterogeneous terminals and services from different manufacturers and providers. It is probable that early 4G terminals will not fully exploit the capabilities of the new system, although they should initially outperform 3G terminals in order to create an attractive market. After launching 4G, it may take several years before we witness advanced terminals taking complete advantage of what broadband communications offer.

References

[1] Alahuhta, P., M. Jurvansuu, and H. Pentikainen, "Roadmap for Network Technologies and Services," TEKES, Finland, 2004.

[2] Karlson, B., et al., "Wireless Foresight-Scenarios of the Mobile World in 2015," 2003.

[3] Frattasi, S., et al., "A Pragmatic Methodology to Design 4G: From the User to the Technology," Proceedings of the 5th IEEE International Conference on Networking (ICN), Reunion Island, France, April 2005, pp. 366–373.

[4] O'Brien, D., and M. Katz, "Optical Wireless Communications Within Fourth-Generation Wireless Systems," Journal of Optical Networking, Vol. 4, No. 6, May 2005, pp. 312-322.

[5] mITF, "Flying Carpet II," 2004.

[6] http://www.arcgroup.org.

[7] http://www.dvb.org.

[8] http://www.dibeg.org.

[9] http://www.3gpp.org.

[10] http://www.worlddab.org.

[11] OMA, "BCASTBOF Report and Recommendation, Version 1.1," January 2004.

[12] DVB, "The Convergence of Broadcast and Telecommunications Platforms," TM2466Rev4.

[13] OMA, MMS Architecture version 1.3, February 2005.

[14] OMA, MMS Requirements version 1.3, February 2005.

[15] OMA-MMSG-2004-0128-MMS-Message-Template-Overview.

[16] OMA, OMA Presence SIMPLE Requirements Version 1.0, January 31, 2005.

[17] OMA, Push-to-Talk over Cellular (PoC) - Architecture version 1.0, February 2005.

[18] Rasmusson, J., et al., "Multimedia in Mobile Phones-The Ongoing Revolution," Ericsson Review, No. 2, 2004.

[19] OMA, DRM Specification Candidate version 2.0, 2004.

[20] OMA, DRM Architecture Candidate version 2.0, 2004.

7

Towards a Unified Convergence on 4G

Throughout the book we have explored and discussed 4G from different perspectives. In this concluding chapter we briefly highlight the most important issues of 4G, summarizing what has been presented in previous chapters and drawing our attention to the prospects for the future, the challenges ahead, and finally some concluding remarks. This chapter will wrap up the journey started in Chapter 1 by presenting a fascinating and universal vision of 4G. We believe that such a vision can lead us towards a viable solution for achieving a true knowledge society as well as to serve individual users while simultaneously being attractive to industry, operators, and service providers.

Chapter 1 gradually introduced 4G by discussing several visions and possible definitions from several sources. In this introductory chapter we could see that there exists several somewhat different conceptions of 4G, though in most cases the visions share some commonalities. Coordinated efforts are needed not only to achieve a worldwide common understanding on what 4G will be, but also to identify and further develop the adequate technical solutions for its realization. In an attempt to provide a sound, rational, and flexible definition, Chapter 2 approached 4G from a user-centric perspective. The only way to guarantee the commercial success of 4G is by first understanding and fulfilling the users' needs and expectations. Chapter 2 discussed user needs and trends in order to identify the user expectations in 4G. Specific and detailed discussions on 4G scenarios and services were discussed. In our vision, 4G can be interpreted as a convergence platform where an eclectic array of networks is supported. The ITU visions and 4G spectrum issues were also presented and discussed in detail. 4G means the convergence of heterogeneous networks, terminals, and services into an everywhere reaching wireless communications system that appears to the user as monolithic and simple. The most promising access techniques for 4G were presented and discussed in detail in Chapter 3. Given the heterogeneity of 4G component networks, it is likely that several access techniques will be used. It is widely agreed that multicarrier modulation and access techniques will play a fundamental role in 4G. The main building block or key enabling technologies of 4G were presented in Chapter 4. These included MIMO technology, radio resource management, software radio communication systems, Mobile IP, and multihop techniques. In order to get better insight on the worldwide research and development towards 4G, Chapter 5 introduced and discussed major initiatives, including the Wireless World Research Forum, the Mobile IT Forum, the Future Technology Universal Radio Environment project, the Next Generation Mobile Communication Forum, the 4G Research Cooperation Projects in the European Sixth Framework Programme, the Worldwide Wireless Initiative, the Samsung 4G Forum, and the eMobility Technology Platform. 4G terminals were considered in Chapter 6 from the technology and application point of view. We identified some trends, potentially good technical solutions as well as challengers for the designers. We believe that, although 4G terminals will come in a large variety of form factors, sizes, and capabilities, a substantial part of these will be multifunctional, multistandard advanced devices. The critical success factors for 4G terminals are service convergence, a user-centric interface, and portable intelligence.

7.1   Short and Long-Term Visions of 4G

In Chapter 2 we described future 4G as a *convergence platform* encompassing and supporting different types of wireless communication networks (see also [1, 2]). We are already living in a period of convergence where *mobile communications, the Internet, computing,* and *broadcasting* are becoming ubiquitous, but also, and equally important, these capabilities are supported by small, portable terminals. Internet and computing, once exclusively linked to large and fixed terminals (e.g., desktop computers), were gradually delivered to smaller devices like laptop and notebook computers. 4G will go one important step further: not only bringing these capabilities to smaller handheld terminals, but also making the connecting wires redundant everywhere, regardless of the terminal type. Also, from a broader perspective, we are witnessing convergence in different areas, as exemplified in Figure 7.1. In fact, *information convergence* has a profound impact on *network, device* (terminal), and *service convergence,* and vice versa. Information convergence refers to the fact that different types of information, in different formats, can be supported by the same system (e.g., transport and access network and terminals). It also refers to the fusing of telecommunications and broadcasting. Service convergence indicates a temporal and spatial continuity in the provision of services, over networks, terminals, and operators.

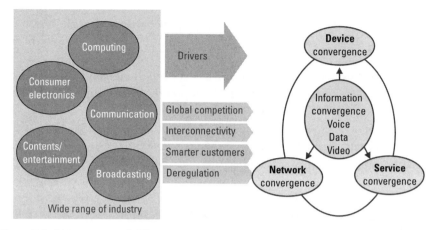

Figure 7.1 Convergence of different worlds towards a ubiquitous wireless communication network.

We see that in the short term the path towards 4G will go through 3G evolution, corresponding to the B3G period, as shown on the roadmap in Figure 7.2 (see also Figure 2.13 in Chapter 2). The B3G technology roadmap can be envisioned either by cellular-based, 2G, 3G, and B3G or nomadic-based, IEEE 802.11 (WLAN), 802.16 (WiMAX and WiBro). 3GPP and 3GPP2 have been active recently in the 3G evolution standards work, and both target the third quarter of 2007 for completion of 3GPP/2 evolution standards. In parallel, the WiMAX camp is also actively stabilizing its IEEE 802.16e specifications (mobile WiMAX) in the second quarter of 2005 and considering its evolution

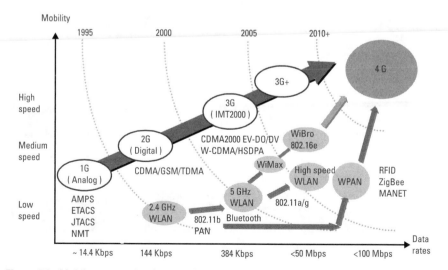

Figure 7.2 Mobile communications roadmap.

standard towards 4G. ITU is actively studying global spectrum allocation for 4G in WRC-07, and with WRC-07 global spectrum results, a long-term 4G standard may start with both evolutionary and revolutionary approaches. Research and business interest in future 4G systems is globally increasing and extensive research in various fields of 4G is progressing through WWRF and regional/national 4G research fora.

New services and killer applications are crucial for differentiating 4G from 3G. For advanced high-speed multimedia services and applications, an IP-centric simplified network architecture is essential to reduce the infrastructure cost. IP networking is also fundamental to provide seamless connectivity in heterogeneous networks. In Korea, the IEEE-802.16e-based WiBro service will start commercially in 2006. We consider 3GPP and 3GPP2 evolution as well as IEEE 802.16e evolution paths towards 4G as a short-term development. In addition, already started parallel developments targeting very high data throughputs for nomadic and mobile environments (e.g., 100 Mbps and 1 Gbps, respectively) based on multicarrier modulation techniques (e.g., OFDM, MC-CDMA) and using a yet-to-be-defined frequency spectrum are seen as the long-term vision of 4G. Figure 7.3 illustrates the evolution of 3G standards towards 4G.

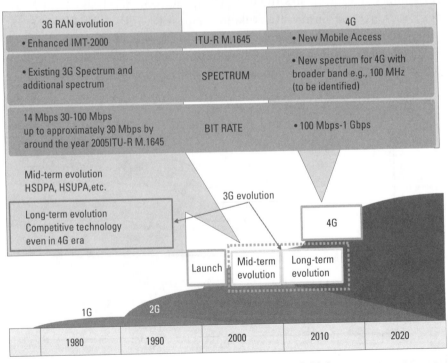

Figure 7.3 Evolution of 3G standards towards 4G (Source: DoCoMo).

7.2   The Challenges Ahead

The idea of a highly interconnected wireless knowledge society is certainly appealing. Although the user is always considered the ultimate beneficiary, terminal and infrastructure manufacturers, operators, and service providers see in this paradigm an unmatched opportunity for their businesses. In its widest sense the 4G concept appears to be the most ambitious ever, and the very first attempt to realize a true knowledge society. The challenges that researchers and developers of 4G will encounter are not few nor are the problems always easy to solve. On the other hand, in terms of social impact and technology development, the reward is well worth the necessary efforts. In what follows, and without trying to provide an exhaustive list, we briefly discuss the key challenges that researchers and developers of 4G systems are likely to face.

- *Multiple 4G definitions.* There is a clear lack of a universal, well-accepted, and unique definition of 4G. Regardless of huge efforts and several years of discussions, there is no worldwide consensus on the definition of 4G. Proposals and visions by ITU and WWRF should be respected and considered as authoritative suggestions for the benefit of all involved partners. Fortunately, developing parties are starting to identify some common characteristics and target capabilities required for 4G systems. Nevertheless, additional efforts should be made in order to ultimately converge towards a common understanding on what is 4G. This will be the first step to guarantee sound foundations and a global footprint for 4G. The importance of a global 4G standard is undeniable for the whole 4G value chain, from manufacturers, carriers, and service providers down to the users.

- *User-centric approach.* Putting the user in the center of the scene appears to be the most logical approach to 4G development. In other words, technology is developed to fulfill users' need and expectations. However, 4G is not necessarily targeting an immediate future and thus, understanding and predicting user needs at a larger time scale is difficult, if not impossible. Moreover, this is time consuming and requires considerable efforts, as there is not just one user profile. Failing to predict the users' needs may lead in the worst case to 4G technology unable to fulfill user expectations. An opposite approach, creating first a generic technology to then trying to match the user needs to that technology is not a better option, in terms of risks. These are somewhat extreme cases, but 4G designers should understand the risk involved. One should not forget also the trendsetter role of industry. It is not only the user who dictates what industry does, industry also create trends that user will follow. It well could be that 4G will emerge from a middle way approach.

- *Seamless interconnectivity.* If 4G is a network of networks, then how do make an array of heterogeneous networks work and appear as a single network regardless of where it is accessed? Seamless vertical (internetwork) and horizontal (intranetwork) handovers are fundamental for achieving ubiquitous connectivity without interruption in the wireless ecosystem. The convergence and integration of existing, evolving, and newly developed technologies is critical to the success of 4G. Since the component parts of 4G (e.g., access schemes for different networks) are being developed independently by different parties, development of interconnection techniques supporting seamless communications should be started at an early stage.

- *Latency.* Several services and applications in 4G will be delay sensitive, particular those exploiting real-time multimedia-based communications. Special care should be put in fulfilling requirements for transmission delays particularly when the contents could be reached from an array of possible networks with possible different access architectures. Connection delays should also be short enough, regardless of the network type or access architecture. Typical maximum accepted values for connection and transmission delays in 4G networks are 500 ms and 50 ms, respectively.

- *Scalability of services in a world of heterogeneous terminals.* 4G terminals will have different capabilities (supported data rate, resolution, processing power, size). How do you efficiently scale up and down services to terminals with different capabilities? This appears to be a difficult task as terminal range is broad and there will be numerous manufacturers and service providers.

- *Exchange of complexity for simplicity.* 4G networks will be complex, but in order to penetrate into the market with success, the user should experience 4G as a simple and intuitive system. Clearly the challenge is hiding networking complexity from users.

- *Technology fragmentation.* By definition 4G encompasses several complementary technologies. As these technologies are developed, they tend to overlap in capabilities, resulting in competition. A severely fragmented 4G will also fragment the market, with negative effects on manufacturers, operators, and users. Technology fragmentation is inherent to 4G, but should lead to complementary systems rather than competing ones.

- *Timetables for 4G.* There is marked difference in the schedule for adopting, deploying, and launching 4G in different regions of the world. Asia has probably the tightest schedule, targeting 2010 as the introductory year. Europe has a less tight schedule, aiming for 2015 or beyond. Early adopters could accelerate the adoption of 4G in other regions, but this might also further fragment or disrupt the worldwide markets.

- *Spectrum.* It is difficult to design a concrete wireless communications system without the knowledge of the frequency band allocated for operation. Knowledge of channel behavior is essential for the system designer. The availability of paired/unpaired bands also has a profound impact on the duplexing mode to be used. Moreover, since 4G assumes the use of MIMO systems, it is difficult if not impossible to design future multiantenna-based systems without knowing the spectrum in which the system will be operating (and hence channel behavior). Spectrum allocation for 4G will hopefully be decided in 2007/2008 by the World Radio Conference. Scarcity of spectrum is another challenge to be taken into account. In addition to the development of communications techniques with higher spectral efficiency, new promising techniques like cognitive radio need to be studied and possibly applied in 4G.

- *Access architectures.* The suitability and limitations of conventional and nonconventional access architectures in different 4G radio networks need to be better understood. Network access architectures include infrastructure (e.g., single and multihop cellular) and infrastructureless (e.g., ad hoc) based concepts. Cooperative techniques are emerging also as a very promising network access approach.

- *Efficient resource allocation.* Management of available resources is crucial to guarantee a truly effective utilization of always-scarce time, frequency, and spatial resources. This is particularly true for 4G systems where multicarrier and multiantenna systems are likely to be used. Finding efficient, fair, and simple-to-implement strategies to allocate resources in multiuser environments is far from trivial.

- *Interference Management.* Multiple access interference in heterogeneous networks with multiples operational air interfaces, heterogeneous terminals and heterogeneous services could be an issue. In addition to the interference issues at network level, also potential interference problems within the terminal should be considered. Indeed, multistandard and multiantenna terminals pose serious challenges for terminal designers, as it could be very difficult to manage and control RF interference within small form factor terminals.

- *Battery technology.* The energy density of current batteries is rather low, and if such batteries were used in 4G terminals they will not provide enough operative time, particularly if rich-content multimedia and high-speed communications are used. Fuel cells have one order of magnitude higher energy densities and so they have the potential to solve this problem. However, fuel cells are still in an immature state. It is expected that the development of fuel cells will reach maturity at around 2010, a timeframe that agrees approximately with the schedule for the launching of 4G in some parts of the world.

- *Power aware design and power management.* Since radical improvements in battery technology are not likely to take place within the introductory period of 4G, power efficient methods and solutions/concepts helping to reduce power consumption in terminals should be critically investigated. This applies basically to the whole design chain and it is related not only to hardware issues like developing low-power consumption components but also it calls for development of power-friendly software and applications. Moreover, techniques, algorithms and system design rules inherently minimizing the power drain will be essential in future 4G systems. Paradoxically, MIMO, identified as a key 4G enabling technology, does not necessarily help to reduce power consumption in terminals. Cooperative techniques, including multihop approaches, used independently or complimentarily with MIMO, are seen as promising to keep power consumption at lower levels.

- *Security issues.* Multimedia-based rich-content communication poses significant security risks for the user, for the information being transferred, and for the content rights. These are fundamental issues that need to be solved before associated services and applications are introduced. Also, since we are considering a highly heterogeneous network where information can be accessed from different places and through different air interfaces, strategies for management of user and information security should be considered.

- *Interlayer design.* Work needs to be done at all OSI layers, separately and jointly. Unlike in previous generations, scarcity and a high demand for resources in 4G leads us to seek efficient solutions to better use the frequency, time, and spatial resources. Cross-layer design and optimization are unique and important techniques that the 4G research and development community is relying on.

- *Costs.* It is expected that the cost of 4G infrastructure should be significantly lower than that corresponding to 3G networks. A target cost of one order of magnitude lower is often mentioned. However, as the prospective band to be allocated to 4G appears to be at higher frequencies than in current 2G and 3G systems, cell coverage is expected to be smaller, leading to the need of more base stations to cover the same area. Also, higher data throughputs shrinks cells significantly. New access architectures, e.g., based on relaying stations, have the potential to solve theses problems. It is clear that lower infrastructure and terminal costs will help to increase 4G network penetration, but considerable research and development efforts need to be done to achieve these goals. Cost per (received) bit of information should also be much lower than in current systems (e.g., one hundred times or even cheaper than at present). Services and their use should be attractively priced. The nature of the information that 4G networks will transfer data requires new pricing strategies.

- *Synergetic efforts.* In order to become a reality, technical challenges have to be solved by worldwide research and development forces. More than ever, concepts and technical solutions should be developed jointly by industry and academia.

- *Potential IPR battlefield.* As 4G represents a mosaic of integrated technologies, it has the potential to become a fertile ground for generating IPRs in various complementary areas. This would benefit the related industry, and also academia. However, as 4G involves some key technologies likely to be present across the whole 4G network (e.g., multicarrier techniques, multiantenna techniques, etc.), some essential IPRs to 4G may have a dominant position, endangering the whole 4G development, deployment and marketing.

- *Open 4G Standards.* In order to succeed 4G should be an open standard and thus, industry should strive to continue with the common practice of 2G, 3G and IEEE wireless standards, among others.

7.3 Conclusions

Convergence is the word that can be best associated with 4G. Convergence can be interpreted in several ways: convergence of wireless and wired networks; convergence of communications, consumer electronics, and computing; and convergence of services, as illustrated by Figure 7.4. From the 4G terminals perspective, one can see a clear trend where several functions and capabilities converge into a single terminal, a multimode (multistandard), multifunction terminal. 4G will materialize the so-called *three-screen convergence*, bringing together TV, PC and mobile phone screens into a single portable device. These screens, perhaps the most notorious technology icons of our time, are watched and scrutinized day after day by billions of people worldwide, mostly at their home, working places and gradually on the move. 4G will not only fuse these separately developed technologies but also and more importantly, it will ultimately free them from any physical connection to the information sources. We

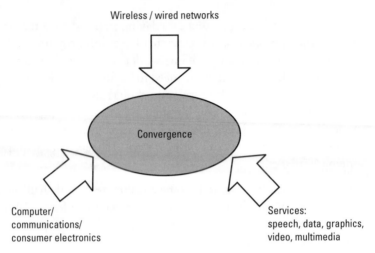

Figure 7.4 Multiple convergence concept in 4G.

call this the *flying screen* concept, as this screen is capable of displaying any content, anywhere the user is, at anytime, and on the move. Figure 7.5 illustrates the three-screen convergence in 4G and the concept of flying screen. Convergence will also take place at the network level, where several heterogeneous networks will appear as to be merged into a single network of networks, as shown in Figure 7.6. Indeed, a 4G network will encompass the entire network hierarchy, from a very wide coverage broadcasting network down to personal networks, and will consider wireless wide area networks, wireless metropolitan networks,

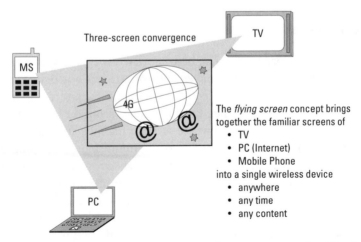

Figure 7.5 "The" Flying screen concept represents the convergence of mobile phones, PC and TV screen.

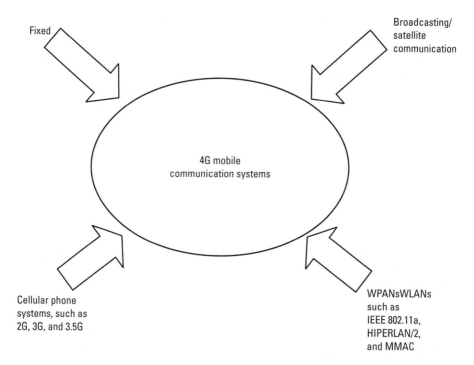

Figure 7.6 Network convergences in 4G.

wireless local area networks, and personal networks as well. To the user this 4G network will appear as a single, simple, and all-reaching network.

As was discussed in Chapter 2, 4G should be approached from the user perspective. 4G cannot be justified just from a technology point of view. Technology, although of fundamental importance, is just the means to satisfy user needs and expectations, and it should come after potentially good services and applications have been identified. Users are not attracted by figures like high data throughput, but they will greatly appreciate useful services exploiting those high data rates. This appears to be the only approach to guarantee an economically successful 4G [2]. The concept of 4G should not only be seen as a technology to provide man-to-man communication but also it should also encompass man-to-machine and machine-to-machine communications.

When one contrasts past and present mobile communication systems with the concepts of 4G, one certainly notices that a paradigm shift is needed in order to better describe and understand future communication systems. Indeed, previous and current systems are basically cellular systems operating in cells with moderate to large sizes. 3G also brought the concept of hotspots providing, for instance, high data rate services mostly in indoor scenarios, as well (vertical) handovers between wide area and local area systems. These ideas are still in 4G systems, but only as component parts. 4G has a wide area mobile component including previous cellular (legacy) systems, their evolution, as well as newly developed very high data throughput systems. In particular, these new systems may not necessarily be based on conventional (single-hop) cellular architectures. Several promising options for access architectures for wide and metropolitan area networks are being taking into account currently, including the concepts of multihop, cooperative and distributed radio. 4G has a fixed or nomadic component for short-range (local area) communications using access architectures with infrastructure (e.g., cellular) or infrastructureless (e.g., ad hoc communications allowing direct or multihop peer-to-peer communications). In terms of coverage, 4G includes everything from personal area networks to distribution (broadcast) networks, and vertical and horizontal handovers work seamlessly, at any place and time. These networks are, in principle, complementary and hence they are designed specifically to coexist. However, at the current development phase, there are several proponent access concepts that clearly overlap in capabilities, and therefore they can be considered as competing. The 4G arena is, by definition, fragmented, but it should not become too fragmented. It is desirable that before the standardization phase begins, a preselection process identifies a reasonable number of potentially good concepts and techniques. We see that WWRF is the most appropriate organization to carry out this job in a global basis.

In an attempt to ultimately define the future 4G network, while taking into account all the observations and existing common understanding, we can state the following:

> 4G is defined as an evolutionary and revolutionary fully IP-based integrated system of systems and network of networks achieved after convergence of wired and wireless networks as well as computers, consumer electronics, and communication technology, as well as several other convergences that will be capable of providing 100 Mbps and 1 Gbps, respectively, in outdoor and indoor environments, with end-to-end QoS and high security, offering any kind of services at any time as per user requirements, anywhere, with seamless interoperability, always on, at an affordable cost, with one billing, and fully personalized.

This is a fascinating vision of 4G. It appears to be the most viable solution for achieving a truly knowledge-based society. It will serve users and still remain attractive to industry, operators and, services providers.

Convergence is really what 4G is all about: convergence of fixed and mobile communications, as well as convergence of regional-area, wide-area, metropolitan-area, local-area, and personal-area networks. 4G is convergence of information, convergence of services, and convergence of terminals. The key is the availability of new technologies. The speed with which telecom operators can roll out converged fixed and mobile services is heavily dependent on the technological revolution taking place at both the network and handset level. Convergence of networks, technologies, applications, and services will offer a personalized and pervasive network to user and will bring a myriad of unrivaled business opportunities to all related parties.

As we have discussed, the key prerequisite to accomplish such a convergence at a technical level is to first achieve a widespread convergence of visions on 4G. We acknowledge the role of WWRF as a key consensus making organization spanning the globe. The sooner the involved parties achieve a common understanding of 4G, and the wider and deeper the consensus, the better the chances of realizing the above 4G definition with technically sound and economically attractive solutions. As the vision imagines, the opportunities that 4G could bring are countless. Opportunities are for the users, society itself, and eventually every single entity, for becoming an integral part of an all-reaching and omnipotent communication network. The opportunities are unmatched for industry and academia to develop and investigate technical solutions. Similarly, operators and service providers will find that 4G is a very fertile land to develop new businesses. Opportunities also mean challenges and risks and they should be identified at early stages.

Convergence is heading towards the advent of a really exciting and somewhat disruptive concept of fourth generation mobile networks. Even though industry and the research community are highly committed to the development of 4G, additional efforts are needed to create a more solid and widely accepted blueprint of 4G. Then, the way towards a smoother and quicker standardization process will be paved.

References

[1]   Prasad, R., "Convergence Paving a Path Towards 4G," *International Workshop in Convergent Technologies IWCT'05*, Plenary Presentation, University of Oulu, Finland, June 6–10, 2005.

[2]   Katz, M., and F. Fitzek, "On the Definition of the Fourth Generation Wireless Communications: The Challenges Ahead," *International Workshop in Convergent Technologies IWCT'05*, Plenary Presentation, University of Oulu, Finland, June 6–10, 2005.

List of Acronyms

3GPP	3G Partnership Project
A4GN	Adaptive 4G Global-Net
AAA	Adaptive Antenna Array
ACF	Auto-Correlation Function
ADC	Analog-to-Digital Converter
AGC	Automatic Gain Control
A-GPS	Assisted-GPS
AHS	Advanced Cruise-Assist Highway System
AMC	Adaptive Modulation and Coding
AP	Access Point
ARP	Address Resolution Protocol
ASIC	Application Specific Integrated Circuit
A-ST-CR	Alamouti-based Space-Time Constellation Rotation
B3G	Beyond 3G
BAN	Body Area Network
BBU	Baseband Signal Processing Unit
BCH	Broadcast Channel

BDMA	Band Division Multiple Access
BER	Bit Error Rate
BLAST	Bell Lab Layered Space Time
BRAN	Broadband Radio Access Network
BS	Base Station
BWA	Broadband Wireless Access
CDMA	Code Division Multiple Access
CIR	Carrier-to-Interference Ratio
CMR	Cooperative Mobile Relaying
CN	Correspondent Node
COA	Care-of Address
CPM	Conference Preparatory Meeting
CR	Cooperative Relaying
CS	Circuit Switched
CSMA	Carrier Sense Multiple Access
CSMA/CA	CSMA with Collision Avoidance
DAB	Digital Audio Broadcasting
DAC	Digital-to-Analog Converter
DARPA	Defense Advanced Research Projects Agency
D-BLAST	Diagonal-BLAST
DECT	Digital Enhanced Cordless Telecommunications
DHA	Dynamic HA
DiL	Direct Link
DL	Downlink
DMFC	Direct Methanol Fuel Cell
DNS	Domain Name System
DRM	Digital Rights Management

DS-CDMA	Direct Sequence CDMA
DSP	Digital Signal Processing
DSPH	Digital Signal Processing Hardware
DSPS	Digital Signal Processing Software
DSRC	Dedicated Short-Range Communication
DVB	Digital Video Broadcasting
EGC	Equal Gain Combining
ETC	Electronic Toll Collection
FC	Frame Cell
FDD	Frequency Division Duplexing
FDFR	Full-Diversity Full-Rate
FDMA	Frequency Division Multiple Access
FFT	Fast Fourier Transform
FH	Frequency Hopping
FP6	European Sixth Framework Program
FPGA	Field Programmable Gate Array
FSS	Fixed Satellite Services
FuTURE	Future Technology Universal Radio Environment
FWA	Fixed Wireless Access
GIMCV	Global Information Multimedia Communication Village
GPRS	General Packet Radio Service
GPS	Global Positioning System
GSM	Global System for Mobile Communications
GW	Local Gateway
GUI	Graphical User Interfaces
HA	Home Agent
HAPS	High Altitude Platforms

HARQ	Hybrid Automatic Repeat Request
HAWAII	Handoff-Aware Wireless Access Internet Infrastructure
HLR	Home Location Register
HSDPA	High-Speed Downlink Packet Access
IC	Integrated Circuit
ICI	Inter-Carrier Interference
IEEE	Institute of Electrical and Electronic Engineers
iDEN	Integrated Digital Enhanced Network
IDFT	Inverse Discrete Fourier Transform
IETF	Internet Engineering Task Force
IF	Intermediate Frequency
IFU	IF Signal Processing Unit
IFFT	Inverse Fast Fourier Transform
IMT-2000	International Mobile Telecommunications 2000
I/O	Input/Output
IP	Internet Protocol
IPR	Intellectual Property Rights
IrDA	Infrared Data Association
IT	Information Technology
ITS	Intelligent Transport System
ITU	International Telecommunication Union
ITU-R	ITU Radiocommunication Sector
ITU-T	ITU Telecommunication Standardization Sector
FA	Foreign Agent
FEC	Forward Error Correction
FN	Foreign Network
FTP	File Transfer Protocol

MA	Mobility Agent
MAC	Medium access control
MAGNET	My Personal Adaptive Global NET
MAI	Multiple Access Interference
MANET	Mobile Ad hoc NETwork
MBMS	Multimedia Broadcast/Multicast Services
MC-CDMA	Multi-carrier CDMA
MIMO	Multiple-Input-Multiple-Output
MIPS	Million Instructions Per Second
mITF	Mobile IT Forum
ML	Maximum Likelihood
MMS	Multimedia Messaging Service
MMSE	Minimum Mean Square Error
MMSEC	Minimum Mean Square Error Combining
MN	Mobile Node
MR	Mobile Relay
MRC	Maximal Ratio Combining
MS	Mobile Station
NAI	Network Access Identifier
NGMC	Next Generation Mobile Communication Forum
NIC	New Industrialized Countries
NLOS	Non-Line-of-Sight
ODMA	Opportunity Driven Multiple Access
OFDM	Orthogonal Frequency Division Multiplexing
OFDMA	Orthogonal Frequency Division Multiple Access
ORM	Orthogonality Restoring Combining
OS	Operating System

OSI	Open System Interconnection
OSIC	Ordered Successive Interference Cancellation
OTA	Over-The-Air
OTDOA	Observed Time Difference of Arrival
OWA	Open Wireless Architecture
PAN	Personal Area Network
PAPR	Peak-to-Average Power Ratio
PCF	Point Coordination Function
PCS	Personal Communications System
PDC	Personal Digital Cellular
PDSN	Packet Data-Serving Node
PER	Packet Error Rate
PHS	Personal Handy Phone System
PIMS	Personal Information Management System
PMR	Private Mobile Radio
PN	Personal Network
PoC	Push-to-Talk over Cellular
PRMA	Packet Reservation Multiple Access
P/S	Parallel-to-Serial
PSK	Phase Shift Keying
QAM	Quadrature Amplitude Modulation
QoS	Quality of Service
QO-STBC	Quasi-Orthogonal-STBC
RA	Radio Assembly
RAB	Radio Access Bearer
RCH	Random Channel
RF	Radio Frequency

RFU	Radio Frequency Signal Processing Unit
RRC	Radio Resource Control
RRM	Radio Resource Management
RSVP	ReSerVation Protocol
RTT	Round-Trip Time
Rx	Receiver
SDMA	Space Division Multiple Access
SF	Spreading Factor
SFH	Slow Frequency Hopping
SDR	Software Defined Radio
SFH	Slow Frequency Hopping
SINR	Signal-to-Interference-plus-Noise Ratio
SIP	Session Initiation Protocol
SMS	Short Message Service
SNR	Signal-to-Noise Ratio
SOC	System-on-Chip
S/P	Serial-to-Parallel
STBC	Space-Time Block Code
STC	Space-Time Coding
STTC	Space-Time Trellis Code
TCP	Transmission Control Protocol
TCU	Transmission Control Unit
TDD	Time Division Duplexing
TDMA	Time Division Multiple Access
TDS-CDMA	Time Division Synchronous CDMA
TeleMIP	Telecommunication Enhanced Mobile IP
TETRA	TErrestrial Trunked RAdio

TFC	Time-Frequency Cell
TFL	Time-Frequency Localized
TH	Time Hopping
TIA	Telecommunications Industry Association
TPU	End-to-End Timing Processing Unit
TRS	Trunked Radio Service
Tx	Transmitter
UHF	Ultra High Frequency
UI	User Interface
UL	Uplink
UMTS	Universal Mobile Telecommunications System
UT	User Terminal
UTRA	Universal Terrestrial Radio Access
UWB	Ultra-Wideband
V-BLAST	Vertical-BLAST
VHDL	Very High Speed Integrated Circuit Hardware Description Language
VICS	Vehicle Information and Communication System
VLR	Visitor Location Register
VoIP	Voice Over IP
VSF-OFCDM	Variable Spreading Factor rthogonal Frequency and Code Division Multiplexing
WCDMA	Wideband CDMA
WiBro	Wireless Broadband
WiFi	Wireless Fidelity
WiMax	Wireless Interoperability for Microwave Access
WINNER	Wireless World Initiative New Radio
WLAN	Wireless Local Area Network

WMAN	Wireless Metropolitan Area Network
WPAN	Wireless Personal Area Network
WRC	World Radiocommunication Conference
WWAN	Wireless Wide Area Networks
WWI	Worldwide Wireless Initiative
WWRF	Wireless World Research Forum
ZF	Zero-Forcing

About the Authors

Dr. Young Kyun Kim is responsible for overall global standards and strategy and research activities within Samsung Electronics. He directs global standards and advanced research on beyond 3G and 4G systems and IP/optical/digital home networking. He is a board member of Open Mobile Alliance (OMA), steering boards of 3GPP PCG/OP, 3GPP2 SC/OP, and Wireless World Research Forum (WWRF).

From December 2000, he has served as a vice chairman of ITU-T SG 19 (Mobile Telecommunication Networks) with interests in mobile core network vision, mobility management and fixed/mobile network convergence. Since 2003, he has served as a chairman of APT Wireless Forum (AWF) and also as a past vice chairman–Asia for WWRF.

Before joining Samsung Electronics, Dr. Kim had 17 years wireless industry standards and research experience in the United States including GTE Corp. and INTELSAT. Most recently, from 1993 to 1999, he served as ITU-R Program Director at INTELSAT, Washington, D.C., representing INTELSAT at the ITU-R SG 8, TG 8/1, SG4, CPM, and WRC.

Dr. Kim has been a senior member of IEEE since 1984 and has published more than 50 papers in IEEE journals, presented more than 70 technical papers at IEEE and other conferences, and served as a regular invited chair and panelist/speaker for many major Beyond 3G/4G conferences and workshops globally. He holds a B.S. from Seoul National University, Seoul, Korea in 1972, an M.S. from Rutgers University, N.J. in 1976, and a Ph.D. from Duke University, N.C., in 1978, all in electrical engineering.

Ramjee Prasad was born in Babhnaur (Gaya), India, on July 1, 1946. He is now a Dutch citizen. He received his B.Sc. (eng.) from the Bihar Institute of

Technology, Sindri, India, and his M.Sc. (eng.) and Ph.D. from Birla Institute of Technology (BIT), Ranchi, India, in 1968, 1970, and 1979, respectively.

He joined BIT as a senior research fellow in 1970 and became an associate professor in 1980. While he was with BIT, he supervised a number of research projects in the area of microwave and plasma engineering. From 1983 to 1988, he was with the University of Dar es Salaam (UDSM), Tanzania, where he became a professor of telecommunications in the Department of Electrical Engineering in 1986. At UDSM, he was responsible for the collaborative project Satellite Communications for Rural Zones with Eindhoven University of Technology, The Netherlands. From February 1988 through May 1999, he was with the Telecommunications and Traffic Control Systems Group at Delft University of Technology (DUT), where he was actively involved in the area of wireless personal and multimedia communications (WPMC). He was the founding head and program director of the Center for Wireless and Personal Communications (CWPC) of International Research Center for Telecommunications–Transmission and Radar (IRCTR). Since June 1999, Dr. Prasad has been with Aalborg University, as the co-director of the Center for Person Kommunikation (CPK) until 2002, and from January 2003 as the Research Director of the Department of Communications Technology and holds the chair of wireless information and multimedia communications. From January 2004 he is the Founding Director of the Center for the TeleInfrastruktur (CTIF) He was involved in the European ACTS project FRAMES (Future Radio Wideband Multiple Access Systems) as a DUT project leader. He is a project leader of several international, industrially funded projects. He is the project coordinator of the European sixth framework integrated project My Personal Adaptive Global NET (MAGNET). He has published more than 500 technical papers, contributed to several books, and has authored, coauthored, and edited 16 books: CDMA for Wireless Personal Communications, Universal Wireless Personal Communications, Wideband CDMA for Third Generation Mobile Communications, OFDM for Wireless Multimedia Communications, Third Generation Mobile Communication Systems, WCDMA: Towards IP Mobility and Mobile Internet, Towards a Global 3G System: Advanced Mobile Communications in Europe, Volumes 1 & 2, IP/ATM Mobile Satellite Networks, Simulation and Software Radio for Mobile Communications, Wireless IP and Building the Mobile Internet, WLANs and WPANs Towards 4G Wireless, Technology Trends in Wireless Communications, Multicarrier Techniques for 4G Mobile Communications, OFDM for Wireless Communication Systems, and Applied Satellite Navigation Using GPS, GALILEO, and Augmentation Systems, all published by Artech House. His current research interests lie in wireless networks, packet communications, multiple access protocols, advanced radio techniques, and multimedia communications.

Dr. Prasad has served as a member of the advisory and program committees of several IEEE international conferences. He has also presented keynote speeches and delivered papers and tutorials on WPMC at various universities, technical institutions, and IEEE conferences. He was also a member of the European cooperation in the scientific and technical research (COST-231) project dealing with the evolution of land mobile radio (including personal) communications as an expert for The Netherlands, and he was a member of the COST-259 project. He was the founder and chairman of the IEEE Vehicular Technology/Communications Society Joint Chapter, Benelux Section, and is now the honorary chairman. In addition, Dr. Prasad is the founder of the IEEE Symposium on Communications and Vehicular Technology (SCVT) in the Benelux, and he was the symposium chairman of SCVT'93.

In addition, Dr. Prasad is the coordinating editor and editor-in-chief of the Springer International Journal on Wireless Personal Communications and a member of the editorial board of other international journals, including the IEEE Communications Magazine and IEE Electronics Communication Engineering Journal. He was the technical program chairman of the PIMRC'94 International Symposium held in The Hague, The Netherlands, on September 19–23, 1994 and also of the Third Communication Theory Mini-Conference in Conjunction with GLOBECOM'94, held in San Francisco, California, on November 27–30, 1994. He was the conference chairman of the fiftieth IEEE Vehicular Technology Conference and the steering committee chairman of the second International Symposium WPMC, both held in Amsterdam, The Netherlands, on September 19–23, 1999. He was the general chairman of WPMC'01 and IWS'2005 which were held in Aalborg, Denmark.

Dr. Prasad is also the founding chairman of the European Center of Excellence in Telecommunications, known as HERMES. He is a fellow of IEE, a fellow of IETE, a senior member of IEEE, a member of The Netherlands Electronics and Radio Society (NERG), and a member of IDA (Engineering Society in Denmark).

Dr. Prasad is the recipient of several international awards and it was worth mentioning that recently he received the "Telenor 2005 Nordic Research Award."

Index

The Artech House Universal Personal Communications Series

Ramjee Prasad, Series Editor

Towards the Wireless Information Society: Systems, Services, and Applications, Ramjee Prasad, editor

Universal Wireless Personal Communications, Ramjee Prasad

WCDMA: Towards IP Mobility and Mobile Internet, Tero Ojanperä and Ramjee Prasad, editors

Wideband CDMA for Third Generation Mobile Communications, Tero Ojanperä and Ramjee Prasad, editors

Wireless Communications Security, Hideki Imai, Mohammad Ghulam Rahman and Kazukuni Kobara

Wireless IP and Building the Mobile Internet, Sudhir Dixit and Ramjee Prasad, editors

WLAN Systems and Wireless IP for Next Generation Communications, Neeli Prasad and Anand Prasad, editors

WLANs and WPANs towards 4G Wireless, Ramjee Prasad and Luis Muñoz

For further information on these and other Artech House titles, including previously considered out-of-print books now available through our In-Print-Forever® (IPF®) program, contact:

Artech House
685 Canton Street
Norwood, MA 02062
Phone: 781-769-9750
Fax: 781-769-6334
e-mail: artech@artechhouse.com

Artech House
46 Gillingham Street
London SW1V 1AH UK
Phone: +44 (0)20 7596-8750
Fax: +44 (0)20 7630-0166
e-mail: artech-uk@artechhouse.com

Find us on the World Wide Web at: www.artechhouse.com